The Field Description of Igneous Rocks

D0843475

R.S. Thorpe and G.C. Brown

Department of Earth Sciences,
The Open University,
Milton Keynes

OPEN UNIVERSITY PRESS
MILTON KEYNES
and
HALSTED PRESS
John Wiley & Sons
New York — Toronto

First published 1985 by
Open University Press
A division of
Open University Educational Enterprises Ltd.,
12 Cofferidge Close
Stony Stratford
Milton Keynes, MK11 1BY
England

British Library Cataloguing in Publication Data

Thorpe, R.S.
  The field description of igneous rocks.
  1. Rocks, Igneous — Identification
  I. Title   II. Brown, G.C.
  552'.1   QE461

ISBN 0–335–10040–6

Published in the U.S.A., Canada and Latin America by
Halstead Press, a Division of John Wiley & Sons, Inc.,
New York

Library of Congress Cataloging in Publication Data

Thorpe, R.S.
  The field description of igneous rocks.

  (Geological Society of London handbook series)
  Bibliography: p.
  Includes index.
  1. Rocks, Igneous — Identification.
  2. Geology — Field work.
  I. Brown, G.C. (Geoff C.)   II. Title
  III. Series.
  QE461.T45 1985   552'.1   84–25182
  ISBN 0–470–20111–8 (Halsted)

Printed in Great Britain

Contents

Acknowledgements

This book would not have been completed without the continued support and enthusiasm of staff at the Open University Press, notably Howard Jones and Rose Dixon. We are particularly grateful to Dr. Mike de Freitas, the Geological Society Handbook Series Editor, and to Professors Peter Baker, Ian Gass, Wallace Pitcher and Dr. Peter Bowden for their detailed and helpful comments on earlier drafts of the manuscript. Last, but not least we thank the following secretarial staff of the Open University for their valuable support during preparation of several draft manuscripts: Ann Budd, Marilyn Leggett and Christine Sanderson.

Illustrative matter appearing in this book was adapted from or inspired by the following sources to whom grateful acknowledgement is also made:-

Fig. 3.4: F.H. Lahee, *Field Geology*, McGraw-Hill (1916); Fig. 3.7: Crown Copyright, British Geological Survey; Fig. 3.9(a): J. Didier, *Granites and their enclaves*, Elsevier (1973); Figs 3.13, 3.14: N.H. Bowden, Liverpool College of Higher Education; Fig. 3.15: C.A. Lee, *J. Geol. Soc. Lond.* **138** (1981); Figs. 4.2, 4.6(a), 4.8–4.11: R.V. Dietrich and B.J. Skinner, *Rocks and Rock Minerals*, Wiley (1979); Figs. 4.4(a), 4.13(c): B. Atkin, Oxford Museum; Fig. 5.2: H. Williams and A.R. McBirney, *Volcanology*, Freeman, Cooper and Co. (1979); Fig. 5.5: P.W. Francis, Open University; Figs. 6.1, 6.6: G.A. Macdonald, *Volcanoes*, Prentice-Hall (1972); Figs. 6.2(a), 6.5: I.G. Gass, Open University; Figs. 6.2(b), (c), 6.9, 6.14, 6.19, 6.20: P.W. Francis, Open University; Fig. 6.7: J.G. Moore and J.G. Schilling, *Contr. Mineral. Petrol.* **41** (1973); Fig. 6.10(b): *Basaltic Volcanism on the Terrestrial Planets*, Pergamon Press (1981); Fig. 6.13: T.R. McGetchin, M. Settle and B.A. Chouet, *J. Geophys. Res.* **79** (1974); Fig 6.17(a) K.H. Wohletz and M.F. Sheridan, *Geol. Soc. Am. Spec. Paper* **180** (1974); Fig. 6.18: G.P.L. Walker, *J. Geol.* **79** (1971) and R.S.J. Sparks and J.V. Wright, *Geol. Soc. Am. Spec. Paper* **180** (1979); Fig. 6.21: R.S.J. Sparks, S. Self and G.P.L. Walker, *Geology* **1** (1973); Table 6.2: R.V. Fisher, *Bull. Geol. Soc. Am.* **72** (1961); Figs. 7.2, 7.3, 7.5: J.B. Wright, Open University; Figs. 8.1–8.4: F.H. Lahee, *Field Geology*, McGraw-Hill (1916); Figs. 8.5, 8.6: M.A. Bussell, W.S. Pitcher and P.A. Wilson, *Can. J. Earth. Sci.* **13** (1976); Fig. 8.9: J.G. Arth, F. Barker, D.E. Peterman and I. Friedman, *J. Petrol.* **19** (1978); Fig. 8.11: R.H. Sillitoe, *Geol. Ass. Canada Spec. Paper* **14** (1976); Fig. 8.12(a): R.V. Dietrich and B.J. Skinner, *Rocks and Rock Minerals*, Wiley (1979); Fig. 8.13: J. Eastwood, *Geology of Cockermouth and Caldbeck*, HMSO (1968); Fig. 9.2: G. Rocci, *Notes Bur. Rech. Géol. Minières, Dakar* **6** (1960); Fig. 9.3: E. Saether, *Kong. Norske. Vidensk. Skr.* **1** (1957); Figs. 10.2, 10.4: C.A. Lee, *J. Geol. Soc. Lond.* **138** (1981); Fig. 10.5: C.A. Lee and M.R. Sharpe, *J. Geol. Soc. Lond.* **138** (1981); Figs. 10.6, 10.8: J.D. Smewing, University College of Wales, Swansea; Fig. 10.9: F.J. Vine, University of East Anglia; Fig. 11.3: J. Duchesne and D. Demaiffe, *Earth Planet. Sci. Lett.* **38** (1978); Fig. 12.5: K.R. Mehnert, *Migmatites and the origin of granitic rocks*, Elsevier (1968).

1

Introduction

1.1 The importance of fieldwork

Fieldwork is the basis of all geological studies, whether they are simply intended to describe igneous rocks by the methods explained in this Handbook, or whether they form the basis of interpretive studies based upon laboratory analysis of rock samples and geochemical and geophysical data collected in the field. Such data *cannot* be interpreted without detailed observations of the field relationships between collected samples and of the geology of areas in which measurements were made. There are many instances in which expensive geochemical and geophysical data have been misinterpreted through incomplete knowledge of basic field relationships. Therefore, *if the appreciation of field geology is poor, then all studies based on collected samples and field measurements will be equally poor*. Conversely, *good appreciation of field geology forms the basis of good geological interpretation*.

Igneous rocks are usually studied first in the field and subsequently in the laboratory. Geological mapping of igneous rocks will involve the methods explained in *Basic Geological Mapping* (Barnes, 1981) but study of igneous rocks might also include petrological and mineralogical investigation, geochemical and isotopic analysis to determine the age and origin of the rock samples, and the use of geo-

physical measurements in the field to determine the distribution of rock-types below the ground. Also, many igneous rocks are associated with distinctive types of economic mineralization and these are always discovered and evaluated by field work. For all of these purposes, *an appreciation of the field characteristics and field relationships of studied samples is essential.* In this handbook we explain how to observe igneous rocks in the field, from the scale of hand specimens and outcrops to that of regional field relationships; these observations provide the foundations upon which most other studies of igneous rocks are based.

1.2 Igneous rocks in relation to regional tectonics

Igneous rocks are materials that have solidified from molten or partially molten material, termed *magma*. Such rocks may be classified as *extrusive rocks*, which were erupted at the surface of the Earth, and *intrusive rocks* which are intrusions that crystallized beneath the surface. In terms of mineral composition, they range from dark-coloured, or *melanocratic* rocks rich in ferromagnesian minerals (e.g. olivine and pyroxenes), through *mesocratic* rocks to light-coloured, or *leucocratic* rocks rich in felsic minerals (e.g. feldspars and quartz). Also, in terms of grain size they range from *fine-grained* rocks, in which individual crystals cannot be distinguished with the naked eye, to rocks in which individual crystals are visible, which are termed *medium-grained* (mean grain size 1–5mm or *coarse-grained* (mean grain size > 5mm). The classification of igneous rocks is discussed in detail in Chapter 4.

Igneous rocks of different compositions and field relationships exist in well-defined associations in which the rocks involved tend to occur in specific regions of the *continental and oceanic crust* and show a distinctive relationship to the surrounding rocks. This reflects the mode of formation and emplacement of igneous rocks in the context of regional tectonic patterns; therefore the following text describes the characteristics of igneous rock associations in relationship to regional tectonics.

The Earth's crust forms the uppermost part of the outer rigid shell, or lithosphere, of the Earth and is divided into large coherent 'plates' that move in relation to one another. Although it is anticipated that most readers will be familiar with the principles of plate tectonics, and although this book is essentially a field guide, a brief summary of plate tectonics provides a helpful framework for understanding the concepts of igneous rock associations and their relationship to one another. However, plate tectonic models are based on global geophysical data and syntheses of many regional geological studies. Regional studies commonly involve areas with dimensions measured in terms of km^2 or tens of km^2, whereas tectonic plates are commonly thousands of km^2 in size. *It may, therefore, be quite impossible to allocate individual small areas of igneous rock unambiguously to a particular plate tectonic setting.*

The boundaries between plates are of four types.

(i) *constructive plate margins* or *ocean ridges*, where the oceanic portions of two plates are moving apart, so permitting the upwelling and solidification of magma to form new oceanic crust.

(ii) *destructive plate margins*, which mark places where two plates are converging so that one plate sinks below the other and is eventually resorbed into the mantle or 'destroyed'. This process is accompanied by formation of a range of magmas within the mantle, on and above the descending lithospheric slab. Such plate margins may occur on oceanic or continental lithosphere forming, respectively, island arcs and active continental margins.

(iii) *conservative plate margins* are faults where two plates slide past each other, so that lithosphere is neither created nor destroyed, and igneous activity is quantitatively minor.

(iv) *collision zones*, where two island arcs and/or continents have collided so that subduction of oceanic material has ceased. Such areas are characterized by widespread extrusive and intrusive igneous activity which commonly continues for a considerable time after collision.

Over 99% by volume of igneous activity occurs at constructive and destructive plate margins and at collision zones and some occurs at locations *within* the plates, for example volcanoes, such as those of Hawaii and those associated with the East African rift system.

Igneous activity may be classified according to its plate tectonic setting as shown in Table 1.1

Table 1.1 Classification of igneous activity in relation to plate tectonic setting

| | PLATE MARGIN | | | | WITHIN-PLATE | |
	Constructive		Destructive	Collision zone	Oceanic	Continental (± rifting)
	Marginal basin (back-arc spreading centre)	Ocean ridge (large ocean basin)	Island arc (gradation to active continental margin)	Continent–continent and continent–island arc collision		
extrusive	Basalt		Basalt-andesite (in island arcs); andesite-dacite-rhyolite (in active continental margins)	Dacite-rhyolite	Basalt-andesite-rhyolite *	
intrusive	Gabbro		Gabbro-diorite (in island arcs); diorite, granodiorite and granite (in active continental margins)	Granite	Gabbro-diorite–granite *	

*These rock types have a distinctly different composition and mineralogy from the analogous rock types formed at plate margins, and a range of rarer alkali-rich rock types may also be found in within-plate settings.

Igneous activity at constructive plate margins is responsible for the formation of the oceanic crust. The composition and structure of the oceanic crust is known from the study of rocks dredged from the ocean floor, from seismic studies which have shown that the crust has a layered structure, and from studies of sequences of older rocks that are believed to be fragments of the oceanic crust. These lines of evidence indicate that the oceanic crust consists of layers of basalt lavas, basalt /dolerite dykes, gabbro and peridotite. These rocks form a distinctive association which may be recognized in ancient orogenic belts, where it is termed the *ophiolite association*. The recognition of such associations is clearly of great palaeogeographic significance and the ophiolite associations are described in detail in Section 10.3.

The oceanic lithosphere moves away from the oceanic ridge by the process of sea-floor spreading and is generally returned to the mantle at a destructive plate margin within *ca.* 200 Ma. The descent of oceanic lithosphere into the mantle is accompanied by partial melting within the descending oceanic crust and overlying mantle to form magmas ranging in composition from basalt, through andesite to rhyolite in composition. These intrude the crust and may be erupted at the surface or emplaced at depth as gabbro, diorite and granite. In some places, the emplacement of such rocks causes melting of the lower crust and this results in the emplacement of intrusions dominantly of diorite, granodiorite and granite composition at destructive continental margins, accompanied by eruption of andesite, dacite and rhyolite. The intrusive rocks emplaced at active continental margins form linear belts of intrusive complexes of diorite-granite composition, termed *batholiths*, which are believed to be underlain by metamorphic rocks. The batholithic association is therefore characteristic of destructive plate margins and is described in Chapter 8.

The composition of the continental crust broadly resembles that of the igneous rocks of andesite composition. Much of continental crust is thought to have formed as a result of igneous activity of the type seen today at island arcs and at destructive continental margins. The crust has evolved continuously as a result of magmatic and metamorphic activity, uplift, erosion and sedimentation, and hence consists largely of metamorphic and igneous rocks with a thin veneer of sedimentary rocks. Because of its greater age and complex geological history, the structure of the continental crust is much more varied than that of the oceanic crust. The continental crust is therefore considered to have a complex structure characterized by rapid lateral and vertical variation, and uplifted sections from which the sedimentary veneer has been eroded expose sections of a wide variety of igneous rocks emplaced at great depth within the crust, including metamorphosed igneous complexes (Chapter 12).

Igneous rocks formed at locations distant from plate margins (locations within-plate, Table 1.1) may have distinctive modes of occurrence, for example, as flat-lying sheets of plateau lavas (Section 5.1), as discordant plutonic magmatic bodies within continental rifts, and as concordant or discordant gabbroic intrusions (Chapter 8). Such igneous rocks may have characteristic compositions; indeed, many magmas emplaced at locations within a plate have distinctive alkali-rich chemical compositions which may be reflected in their mineralogy (Chapter 9 describes the alkaline intrusive association).

The organization of this field guide emphasizes the occurrence of igneous rock associations within characteristic crustal regions. Table 1.1 provides a summary of the relationship between igneous rock associations and plate tectonics. Chapters 2–5 deal with the approach to making field observations in igneous terrain, and the interpretation of such observations is introduced in Chapter 5. Finally, Chapters 6-12 consider the characteristic features of igneous rock associations in a regional geological and tectonic context.

1.3 Geological interpretation of field work

The study of igneous rocks in the field will involve different strategies. For example, field study may be pursued 'objectively' — observations are recorded impartially so that, after the field work, an interpretation can be made on the basis of the collected evidence. Alternatively, a geological study may be planned with a particular aim in mind; for example, the search for economic mineral deposits, or the collection of samples of a particular rock type. In such cases it is quite natural to carry out field work with the aim of searching for evidence of mineralization, looking for particular rock types or gaining evidence for, or against, a particular hypothesis. While avoiding a completely subjective approach, field work may quite legitimately be planned to 'prove' or 'disprove' a particular hypothesis. These two strategies are complement-

ary, and in practice should be accompanied by formulation and revision of working hypotheses. When carrying out fieldwork (and other geological activities) you should bear in mind the philosophy of this quotation from the mining geologist Ira B. Joralemom :

> The geologist must not devote himself to any one theory or to the facts that support it long enough to fall hopelessly in love with it. He must make each theory the object of a summer flirtation and not a wife — and he must be ready to throw each one over the moment a more attractive mental maiden comes along. (Ira B. Joralemom (1940), Mining Congress Journal pp.35–6)

How this occurs will depend upon personal preference and experience (which *cannot* be gained from a book or a university course!). So, during field work, you should think continually about the significance of field observations without adopting extreme strategies; *note that the mere recording of field observations without consideration of their significance is almost as bad as reconciling field data to a preconceived hypothesis.* Nevertheless, accurate data will have a permanent value even if their significance is not appreciated in the field.

Remember that the approach you adopt and the interpretation of field data that you make will depend upon your experience, and reflect your knowledge of regional geology and rock associations. Interpretation of field observations requires a knowledge of field associations but *knowledge of such associations depends on geological fieldwork.* Such work is based on careful systematic field study

of rocks, and this handbook describes
how to approach the study of igneous
rocks in the field.

2

Field techniques

The techniques used for a particular project will clearly depend upon the aims and objectives of the project. These might include mapping on a large-scale reconnaissance survey (measured in hundreds of km²) allied with selected field traverses, or detailed mapping of a small area (perhaps a few km²) based upon particular description of individual outcrops. For some projects a published geological map of an area may be used to collect samples from different rock units for thin section study and/or chemical analysis. Although these aims and objectives are clearly varied, it is emphasized that *all geological work* is based upon *accurate geological maps*, and that observations (such as thin section descriptions, mineralogical and chemical analyses) *cannot be interpreted* without a *careful description of the field occurrence within the context of an accurate geological map.*

2.1 Preliminary studies

Suppose that it is necessary to produce, from first principles, a reconnaissance, regional or detailed geological map (cf. Barnes, 1981, Chapter 3). In such circumstances, it is helpful to make a planning visit to the field area for a few days *well before commencing* a detailed study that may take several weeks or months. The nature of the mapping carried out, or the maps

to be used, determines field practice; for instance, accurate delineation of the distribution and volume of volcanic formations might require detailed field mapping, and measurement of stratigraphic sections. For intrusive rocks, mapping of complex contact relationships may necessitate mapping of individual outcrops on a scale of 1 : 10,000 (*ca.* 6 inches to the mile), or occasionally at larger scales (e.g. 1 : 2500, *ca.* 25 inches to the mile). In both cases, the preliminary visit will allow the broad distribution of the major rock-types present in an area to be assessed.

For all studies it is important to obtain a detailed topographic map, aerial photographs and, where possible, a published geological map. When a published geological map of an area is used, a reference should be included in any written and/or published work stating exactly how it was used. Note that published maps may not be perfect and you should always evaluate the extent to which your field observations support or contradict a published interpretation; in some cases time may be wasted in attempting to justify a published map when a different interpretation might be more appropriate. Available maps and photographs should be studied *before* going into the field (and copies should be taken into the field) as an aid in formulating the field plan; for example, in saving time crossing ground where there is little chance of finding good outcrops.

For field studies involving collection of samples for geochemical analysis, it is helpful to transfer details of lithological boundaries, major outcrops and quarry sites from the geological map (if published) onto the topographic map and/or aerial photograph. Where outcrops cannot be identified (or where no map is available), then much time may be required to find suitable exposures, especially in glaciated lowland areas. Obvious places to start looking are sea cliffs (on coastal areas!), stream sections, road and railway cuttings and construction sites, especially for large engineering works such as dams and tunnels. Quarries generally provide excellent sections, but the owners should be contacted *before arrival* to check that access will be possible. Permission should *always* be obtained to examine exposures on private land.

After the exposures have been located, a decision should be taken on how many should be examined, and in what degree of detail. Again, this will depend upon the aims of the project as well as upon time, access and weather. On a reconnaissance survey a small number of well-spaced sections should be studied but, if a specific rock-type (such as a volcanic unit or an igneous intrusion) is being mapped, then there is no short-cut and all available outcrops should be examined, with more attention being given to outcrops likely to show important details.

Outcrops should first be viewed from a distance to determine any gross structural features which should be assessed to determine whether they relate to the mode of occurrence of an igneous body (cf. Chapter 3). The outcrop should then be studied more closely, to see what smaller scale variations occur, and to decide whether such features are worth describing in detail. For *all* outcrops visited it is important to note their main geological characteristics, including the overall features of the outcrop and to make notes on samples collected (details in Chapters 3 and 4).

2.2 Equipment

The equipment required for a particular project depends upon the detailed aims of the work. For almost all projects, *a hammer* (ca. *1 kg*), *a notebook (with adequate pens, pencils and erasers), compass/clinometer, steel tape/ruler and handlens are essential.* A camera is invaluable and, when samples are to be collected, a *heavy hammer* (ca. *2–3 kg*) *with a strong shaft, a waterproof marker and polythene or cloth sample bags should be carried.* For detailed geological mapping, a *long steel tape* (e.g. 20–30 m) is particularly valuable and, for large-scale studies, a stereoscope for the study of aerial photographs may prove useful. A geologist must be equipped (everywhere) with good footwear and with clothes appropriate for the climatic conditions. Finally, a small *first aid kit* should always be carried.

The detailed specification of field equipment has been described in Barnes (1981, Chapter 2), where the reader will find a description of the different sorts of hammers (Section 2.1), compasses and clinometers (Section 2.2), handlenses (Section 2.3), tapes (Section 2.4), map cases (Section 2.5) and field notebooks (Section 2.6) that are commonly available. *Safety spectacles* should be worn when breaking samples from hard rock. For most studies of igneous rocks, the *collection of samples* is important; even for straightforward geological mapping, small samples should be collected and labelled for comparison with rocks from other outcrops (cf. Chapters 3 and 4). When collecting large rock samples for thin section study and chemical analysis, the weathered surface should be removed as far as possible; it is useful to carry a smaller chisel-ended geological hammer for this purpose.

2.3 Field notes

Although it may take some experience to develop a facility for taking notes at an appropriate level of detail, to start with it is best to record *as much detail as possible*. If this is done, then you will have a complete set of notes, and with experience it will become relatively easy to refine the technique of taking field notes in the appropriate detail and style for the project; *too much detail is always preferable to too little detail in field notes*. There is nothing more frustrating than returning from remote, inaccessible field terrain only to discover a lack of detail in some aspects of the field notes, coupled with a hazy memory. Incomplete notes can render worthless all the work done at a location and the effort expended in going to remote exposures.

Since most of the results of field work should be recorded in the field, *the field notebook and field map provide the most important record of field observations and collected specimens. The field notebook and field map are therefore vital pieces of equipment which should be carefully written and preserved.* Always remember to write your name and address in the front of the notebook. Your notebooks are so important that, in many cases, it is useful to transfer details of outcrops visited and specimens collected, to a separate notebook at the end of each field day, so as to avoid loss of vital data if the notebook were mislaid. Whether carrying out reconnaissance mapping or sample collection based on a published map, each outcrop should be identified, described and numbered in the field notebook. In poorly exposed areas, it is relatively easy to identify, and number an individual outcrop. This may be marking it on an aerial photograph and/or map, and recording it in the field notebook; for example, 'small hills of leucocratic volcanic rock, *ca.* 100 m in diameter, 100 m southeast of disused windmill on top of Parys Mountain, Anglesey.' In a terrain of continuous exposure small areas of outcrop should be identified and located on the map, using compass methods where necessary. Samples collected should also be located on the map and described in the field notebook (see also Section 2.4). Finally, it is especially important to describe the outcrop sufficiently well that it may be located independently by another investigator who may wish to collect rocks to confirm your observations or to extend them in other directions. An example of the description and sample collection from an individual outcrop of igneous rock is shown in Fig. 2.1.

Field notes should be factual and describe exactly what is being observed. They may include interpretations or speculations made (or imagined) in the field but these should be *clearly distinguished* from direct field *observations*. It is important to describe the size, shape and orientation (e.g. dip and strike) of any features particular to the igneous rocks that you observe. This should be done with accurate labelled sketches (or photographs) which should *have a scale*. Record the location and subject of photographs in the notebook.

(a) *Thursday 10 June 1981*

Moel-y-Penmaen; small polygonal hill ca. 300×150m, at 300m NNE of Penmaen farm (G.R. SH338387) (cf. Mately and Heard, 1930, plate 16).

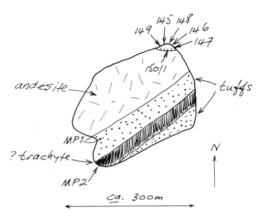

The hill is composed of NE-SW striking pyroclastic rocks and lavas; south side of outcrop contains fine-medium grained poorly-bedded tuffs and grey aphyric lava (? trachyte; MP2). Central part has medium-grained pyroclastic rock - also poorly-bedded (sample MP1). The northern part of the hill is composed of massive grey, flinty plagioclase-phyric andesite lava. Samples LL145-151 (7 samples) collected from disused quarry working in NE corner of hill; 145-148 on or up to ca. 1m above floor level, 150 = ca. 4 m above base and 151 = ca. 6 m above base.

Fig. 2.1 Examples of description and sample collection from outcrops of igneous rock from (a) the Palaeozoic of the UK and (b) the Tertiary of NW Argentina. Note the use of sketch maps and sections, the record of dimensions and the use of abbreviations, for example, v.f.g. for very fine-grained (see Barnes, 1981).

(b) Sunday 27 December 1981

Section of volcanic rocks exposed in river section
ca. 3 km north of houses at Trapiche (exact
location marked on aerial photograph).
The cliff on the east side of the valley is composed
of homogeneous light grey andesite/dacite lavas
with sparse plagioclase phenocrysts. These lavas
are slightly vesiculated and have a blocky/platey
jointing pattern and are over lain by a flow of
dark-coloured olivine-plagioclase basalt/basaltic
andesite. The contact is sharp and well-defined
and marked by ca. 1m of rubbly vesicular
basaltic blocks above weathered surface of
andesite flow. The lavas are overlain by a
crystal-rich, pumice-poor ignimbrite containing
feldspar, quartz, biotite and Fe-Ti oxide in a
fine-grained matrix. Ignimbrite has a low
concentration of lithic and pumice fragments
(maximum lithic fragment = ca. 0.5 cm) and is
either non-welded a poorly-welded. No internal
divisions within the ignimbrite were observed here.
The west side of the valley is a younger? dacite dome
with platey joints which are locally variable in
trend but generally dip towards the valley,
defining the external form of the dome. Contact
with the older ? lavas is not exposed but is
assumed to be within alluvium (rubble within the
valley).

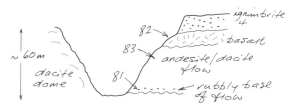

81 block from base of andesite cliff alt. 3980 m.
plag-phyric andesite with distinctive speckled
texture.
82 basalt outcrop alt. 4020 v.f.g. tiny phenocrysts of
ol + plag in matrix.
83 andesite. Rubbly top of flow is exposed, weathered
reddish below basalt. Lava has good flow banding,
defining phenocryst rich and - poor zones. Sample
is hb (?)-ol (?) plag-phyric with g/mass speckled
as in 81.

11

2.4 Collecting specimens

Specimens should generally be collected from exposures *in situ*, unless there is *no reasonable doubt* about the location of a block that is not *in situ*, for example, where a large number of identical blocks lie at the bottom of an inaccessible crag. If the specimen is required *only* for a thin section and comparison with other outcrops, then a relatively small fresh piece may be collected, perhaps only 200–300 g. When there is a directional fabric in the rock, then the orientation of the specimen (dip and strike, and an indication of the top and bottom) should be recorded *on* the specimen. Where the specimen is to be analysed chemically, a larger piece should be collected; for many purposes, a fist-sized specimen of fresh rock (*ca.* 1–2 kg) is adequate. For some purposes, for example collection of samples for mineral separation or isotopic dating, larger samples (2–3 kg) may be necessary. In statistical terms, a *representative* sample of a coarse-grained rock such as a granite should be *much larger* than that of a fine-grained rock, or a glassy rock. Although it is desirable to collect large specimens of coarse-grained rocks, the size of samples collected may be limited by practical considerations, such as transport.

For specimens collected in the field, *accurate labelling with their field numbers is of vital importance. Specimens with illegible or indecipherable numbers are useless.* For most rock collections, the specimens should be labelled at least once with a black water-resistant felt-tip pen (e.g. *Suremark Magnum Marker* or *Pentel* pen) and placed in a tough plastic bag (also marked, with the field number), or wrapped in newspaper. To prevent field numbers from being abraded or obscured during transport in a rucksack (or elsewhere), it is often worth writing the number on a piece of paper to be placed in the specimen bag as a safety precaution. As for field notes, locate the specimen on a map or aerial photograph and enter a brief description in the field notebook. This should be adequate to distinguish the specimen from others and will prevent any confusion of specimen identity; for example, 'light grey plagioclase-, hornblende-(?) phyric andesite showing some alteration to chlorite and epidote along veins (2 pieces)' (see Chapter 4 for more details).

2.5 The scale of observations

The preparation will start with examination of maps and/or aerial photographs *before* visiting the field, and may include a short preliminary visit (2–3 days) to the field area coupled with reconnaissance sampling. Following these preliminaries, the remainder of the field studies will involve progressing through different scales of observations, starting with those on hand specimens used to interpret individual outcrops, and passing on to the use of outcrop information to interpret the geology of a larger area. This increase in scale of observations is implicit in many geological studies, as explained in Chapters 3 and 4.

When preliminary studies have been completed, the task of working with igneous rocks in the field begins in earnest. At this stage, it is tempting to rush up to an outcrop in order to identify rock-types immediately. However, a more cautious approach is to be recommended: stand back from each outcrop and note its broad features before making more detailed observations. In particular, look for clues about the mode of occurrence and nature of the igneous body that may be inferred from the features of individual outcrops. First, record in the field notebook your location, allocate an outcrop number and comment on the general form of the outcrop viewed from about 10–20 m distance. For example: 'Bluebell Farm (beneath rear wall); G.R. 462381; outcrop number 84-01-05 (using year, month, outcrop number); massive homogeneous face of dark pink-brown rock, *ca*. 10 m wide by 1.5 m high, ice-smoothed, vertically jointed — irregular joint planes *ca*. 20 cm average separation.' In this way you will note all distinctive features that may be present, such as jointing, veining, contacts, flow banding, occurrence of xenoliths and so on, before making close-up studies and collecting samples. If the outcrop does show any of these large-scale features, then it is good practice to make a field sketch and possibly to back this up with a photographic record. Do not forget to indicate the direction of view, to use a scale and to record the outcrop number on the field sketch and list of photographs. (Note that some of the terms defined in Tables 4.2, 4.3 and 4.4 are used in this Chapter; forward reference may be necessary if unfamiliar terms are encountered.)

3.1 Jointing

Usually the most obvious features of an igneous rock outcrop are several sub-parallel sets of cracks or fractures, varying from a few centimetres to a few metres apart, known as *joints*. The face of the granite quarry in Fig. 3.1, for example, has three prominent joint sets, two that are near vertical (one set parallel to, and one set at right angles to, the main face) and one that is horizontal. The origins of joints in intrusive igneous bodies are complex, but may be attributed to (i) the contraction that occurs when a volume of magma crystallizes, producing joints that may bear a simple geometrical relationship to the intrusion walls and which may, therefore, be useful in deducing the shape of the body (Chapter 5); (ii) post-intrusive regional tectonic stress, producing joints that usually extend beyond the intrusive body and may be matched to joint patterns in the surrounding country rocks; and (iii) expansion due to the removal of the confining pressure of overburden — this will widen pre-existing joint sets, and possibly increase their numerical density, or may create new sets with different orientations. Joints in minor intrusions and extrusive igneous bodies that have not been buried significantly and have not experienced deformation or faulting will usually relate simply to the tension created by cooling and contraction. *When mapping the orientation, i.e. dip and strike, Fig. 3.2 (Barnes, Chapter 5), note the average spacing of joint sets measured at each outcrop for later interpretation by comparison with other outcrops.*

Fig. 3.1 Granite quarry face at Shap, Cumbria, showing strong vertical jointing parallel to and at right angles to the face and a more widely spaced horizontal joint set. Width of field of view = 30 m.

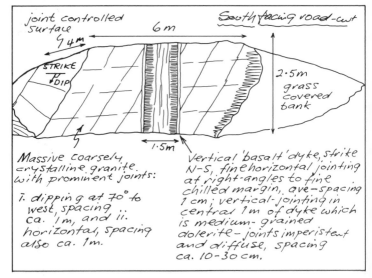

Fig. 3.2 Field sketch of locality showing jointed granite cut by a vertical dyke.

14

Faults are surfaces, sometimes joints, upon which relative motion of opposite sides has occurred, and may not be easily identified in homogeneous igneous rocks where the two displaced sides cannot be 'matched'. However, the presence of *slickensides* (grooves or striations on the sides of a fault, parallel to the direction of movement, Fig. 3.3) and of crushed or fragmented *brecciated* rock along the joint should be noted, as both are evidence of faulting. Such brecciation may result in formation of a cemented *fault-breccia* and in some cases, especially where softer rocks are involved, the rocks may be crushed to a soft, uncemented clay-like material, called a *fault-gouge*.

3.2 Veins and contacts

Veins are sheet-like, or tabular bodies that occur *within* both igneous and country rocks.

Hydrothermal veins are white and usually of quartz or calcite. They vary in width from a few millimetres to a metre and are very common in igneous terrain. The mineral constituent may be distinguished as quartz or calcite using the hardness and cleavage tests described later (Table 4.6). Such vein materials may be deposited either in pre-existing joints or in newly-developed joints. If the igneous host is cold, the veins will tend to be straight-sided, whereas reaction between vein material and hot, recently-crystallized igneous rocks results in veins with irregular and diffuse edges. Hydro-thermal veins are usually very much longer than they are wide, typically by a factor of 100–1000, and they often vary in thickness and may pinch, swell, branch and swerve; they may also lie at any angle from vertical to horizontal. A collection of dip and strike measurements on hydrothermal veins,

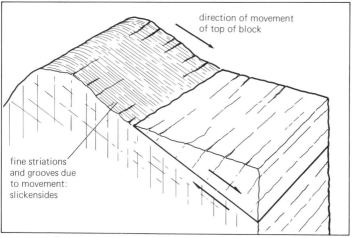

direction of movement of top of block

fine striations and grooves due to movement: slickensides

Fig. 3.3 Block diagram illustrating the development of fine striations, or slickensides on a fault plane.

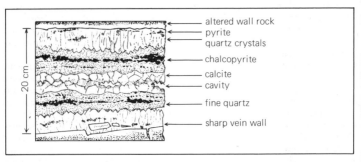

Fig. 3.4 Sketch cross-section across a complex hydrothermal mineral vein.

together with observations on their lateral continuity, may provide useful information on regional stress fields and on the relationship of hydrothermal veining to an exposed (or concealed) intrusive body that provided the heat to promote fluid flow.

Occasionally, complex veins may be found. These contain suites of different minerals arranged in bands (see Fig. 3.4) with a definite sequence from the walls to the centre of the vein. The variations in mineral content reflect the changing species that were being precipitated, as time progressed, from the hot fluids permeating through the vein. Successive linings, usually of lower temperature hydrothermal minerals, are produced (the sequence quartz → calcite is typical) and, at the centre, there may remain an open cavity. Metallic sulphides, such as iron sulphide (pyrite), copper-iron sulphide (chalcopyrite) and lead sulphide (galena) may form minor constituents of mineral veins. A field sketch of a complex mineral vein, as in Fig. 3.4, is an excellent way of recording the fine detail.

Hydrothermal fluids permeating through joints are also often respon-sible for the alteration of the host rocks on either side, involving, for example, chemical reactions that break down feldspars to softer minerals such as clays. If little or no material is precipitated along the joints, wall-rock alteration may be the only sign of past hydrothermal processes (e.g. the dark *greisen* — secondary mica plus chlorite with quartz, derived from granite — veins in Fig. 3.5).

Fig. 3.5 Dark greisen zones in granite at Cligga Head, N. Cornwall. The greisen was formed by wall-rock alteration of host granite as hydrothermal fluids passed through joints along the axes of the dark zones. Average width of greisen zones is about 10 cm.

16

Quartz, calcite and, sometimes, other carbonate veins are precipitated from hot water by post-magmatic hydrothermal processes: two other kinds of late-stage magmatic veins are also encountered frequently in igneous intrusions: aplites and pegmatites.

Aplites are veins of fine-grained felsic (i.e. generally light-coloured, *see* p. 28) crystalline material, usually of a paler colour (white or pinkish) than their hosts (Fig. 3.6). Aplites probably represent a residual fraction of relatively silica-rich magma which remains within the intrusion when the bulk of the magma has crystallized. If this magma is then injected into the solidified parts of the intrusion, it will cool rapidly and form the fine-grained veins dominated by quartz and feldspar crystals seen in aplites. Aplites are

particularly common in granitoid intrusions where, to judge from their lack of chilled margins (see below), they are usually intruded into hot solidified material.

Pegmatites are veins of exceptionally coarse-grained, igneous-textured material (Fig. 3.7), usually incorporating the same minerals as their hosts, that is, quartz, feldspars, micas and occasionally amphiboles. Some may incorporate occasional large crystals of exotic minerals such as apatite, tourmaline (Fig. 3.8; Table 4.6); green beryl $(Be_3Al_2Si_6O_{18})$ and yellow to lilac spodumene $(LiAlSi_2O_6)$. Like aplites, pegmatites are thought to represent the residual melt of a crystallized intrusion, but, unlike aplites, the silicate liquids that form pegmatites are often rich in water and other volatiles

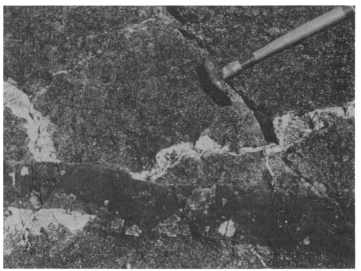

Fig. 3.6 Veins of leucocratic aplite and melanocratic fine-grained gabbro cutting coarse-grained gabbro in the Cuillin complex, Isle of Skye, Scotland. Age relations deduced from cross-cutting relationships are discussed on p.21. Length of hammer is 35 cm.

(e.g. fluorine and boron) and rare trace elements (hence the exotic minerals noted). These volatiles are responsible for depressing the freezing temperature of the pegmatite liquids, so allowing the characteristic large crystals to form (these range from 1 cm to as much as several metres in length). Simple, homogeneous igneous-textured pegmatites without mineral zonation are termed *simple pegmatites*; those with mineral zonation along the length of the vein, and often with exotic minerals, are *complex pegmatites*.

Fig. 3.7 A hydrothermal quartz vein containing black wolfram crystals at the margins at Castle-an-Dinas wolfram mine, Cornwall. The vein is 60 cm wide and intrudes thermally metamorphosed shales. (BGS photo)

Hydrothermal quartz veins, aplites and pegmatites are commonly intimately associated, sometimes occupying the same complex fracture with multiple infillings and sometimes cutting across one another. They are found most frequently in granite intrusions because the magmas that form these rocks are usually rich in the volatiles that concentrate in residual liquids and enhance their mobility. At the stage of hydrothermal quartz vein-aplite-pegmatite formation, the outer skin of the hot intrusion will have cooled to form a rigid surface that may, together with the adjacent country rocks, become cracked and jointed under the influence of pressure variations in the cooling magma chamber beneath. It is these interconnected cracks and joints that become filled by late-stage liquids which may crystallize to form a *stockwork* of veins and in some cases (e.g. the Southwest England granite Cu-Sn mining area) these veins may be

Fig. 3.8 Pegmatite in granite from Luxulyan, Cornwall, showing development of black acicular (needle-like) tourmaline crystals. Area of photograph is 25 × 15 cm.

hosts for economic metalliferous mineralization. Since it is common for the veins in such a stockwork to increase in number towards a contact zone, careful mapping of vein densities provides a useful field guide when attempting to locate the contact between an igneous intrusion and its country rock envelope.

It will be clear that both aplites and pegmatites may extend beyond the boundaries of their host intrusion as veins cutting country rocks. This means that the veins may be sheet-like intrusions which are discordant to the pre-existing fabric of their surrounding rocks. Such intrusions are termed *dykes* and, where they are about a metre or so in width, they might be termed an *aplite dyke* or *pegmatite dyke*. (NB In cross-section, a dyke is a straight-sided vein. The more common forms of dyke are described together with other minor intrusions in Section 7.1.)

Fig. 3.9a Rounded, joint-bounded blocks of gabbro invaded and net-veined by granite at Le Peyron, Massif Central, France. (Jean Didier, 1973.) Length of hammer is 30 cm.

Aplites and pegmatites are usually assumed to have formed at a late stage in the crystallization of the same magma as their hosts. However, at the scale of an outcrop, one sometimes finds that a new generation of magma, perhaps due to renewed magmatism at the site of an old intrusion, has cut slightly older but related igneous rocks. A particular example of this, not associated with pegmatites or aplites, might be where a granitic magma has invaded a rectangular joint set in a previous mafic (i.e. generally dark-coloured, see p. 28) igneous body (Fig. 3.9a). This is known as *net-veining*. As mafic magmas generally crystallize at higher temperatures than felsic magmas, a stage in the crystallization of a mafic magma body may be reached at which it causes local melting of the host rocks to produce granitic liquid. This liquid may then penetrate joints within the cooling mafic body and its surrounding — a process known as *back-veining*. (Fig. 3.9b).

In cases where hot igneous material has been brought into contact with cold host rock, a relatively fine-grained *chilled margin* will often form inside the contact. The nature of the contact and the width of the chilled margin both depend on the temperature difference between an intrusive magma and its host which, in turn, reflects factors such as the original depth of magma emplacement and the composition of the magma (since basic magmas crystallize at higher temperatures than acid magmas). Most intrusions emplaced into hot country rocks will show little sign of marginal chilling, but the contact may have undergone plastic deformation, it may be diffuse due to remelting across the boundary, or it may be disrupted if magma has penetrated into thermally

19

Fig. 3.9b Back-veining of Coire Uaigneich granite into Tertiary basalt due to heating from associated Cuillin gabbro: Camasunary Bay, Skye, Scotland. Area of picture is *ca.* 40 × 25 cm.

metamorphosed wall rocks. In such cases, the material on either side of the contact will be crystalline, although the contact may be recognized through contrasts of grain size, colour and mineralogy. At the opposite extreme, intrusions emplaced into cooler country rocks at a high level within the crust frequently exhibit chilled margins (Fig. 3.10) of rapidly cooled magma, ranging from a few millimetres to a few metres in width, abutting country rocks which may show obvious signs of contact metamorphism (see Section 8.5 for further details). In the special case of net-vein and back-vein complexes (Fig. 3.9), the light and dark rocks sometimes show irregular contact relationships, indicating the simultaneous emplacement of immiscible felsic and mafic magmas. In these cases, the mafic material,

which usually freezes first, may show chilling against the felsic segregations that remained fluid until later in the history of crystallization.

In general, when describing a contact zone involving igneous rocks, measure the orientation of the contact in two dimensions, and, where possible, in three dimensions, to help you trace the feature from one outcrop to the next. Make notes on the orientation, rock-types and geometry of veins, aplites, pegmatites and chilled margins, all of which are closely related to physical processes occurring at or near the contact of an igneous body with its host. Note also any other features of contact zones, such as the occurrence of country rock inclusions (xenoliths) and contact metamorphism in the surrounding rocks, and try to develop an overall picture of contact relations, using as a guide Sections 5.1 and 5.2 on the geometry of igneous rock units.

Fig. 3.10 Contact between coarse-grained gabbro (top) and fine-grained gabbro (bottom) at which the latter shows a thin (*ca.* 1 cm thick), jointed, chilled margin of basalt indicating that it intruded the coarse-grained gabbro. Fine veining in the lower half of the picture is due to quartz veining. Cuillin complex, Isle of Skye, Scotland. Area of picture is *ca.* 8 × 5 cm.

Finally, if an outcrop contains cross-cutting sets of joints, veins, dykes and other contacts, then it can be immensely useful to establish a sequence of geological events by examining their relationships. For example, from Fig. 3.6 it is clear that the coarse-grained host rock (gabbro) was cut first by the light-coloured aplite veins with diffuse margins, followed by the dark-coloured, finer-grained vein (in fact, a dyke) which has chilled margins against the gabbro. A subsequent NE–SW joint set (lower left) cuts both the dyke and aplite which appear to have more brittle fracturing characteristics than the gabbro. Less certainly, a NW–SE joint set (upper right and middle left aplite) may be much older if, as appears possible, the NW–SE-trending aplite vein formed within one of these joints.

3.3 Flow banding and igneous lamination

These features, again easily recognized at outcrop scale, may originate when a magma is partially crystallized and consists of both crystals and melt. Banding may also result from preferential vesiculation (i.e. the development of gas bubbles) along flow lines of a lava. For example, *flow-banded rhyolite* lavas (Fig. 3.11) with parallel bands of slightly different colour, on a centimetre or millimetre scale, result from the streaking out of glassy or partially devitrified layers during movements of viscous lava. Parallel alignment of early-formed minerals in less viscous lavas of intermediate and mafic compositions may also be expressed as banding parallel to the direction of movement, and it is typical of feldspars in some types of andesites and basalts.

On a larger scale, intrusions of all kinds may develop flow structures imparted either by migration of magma from its source to its zone of emplacement, or by subsequent convection within a chamber before the magma consolidated. In granitic rocks the parallel alignment of large well-shaped feldspar phenocrysts (Fig. 3.12), sometimes termed megacrysts (see below), is particularly common near the margins of intrusions where flow-aligned inclusions may also occur. It will be self-evident that, where *preferred orientation* of minerals or inclusions occurs in an intrusion, several faces of an exposure must be studied at the same outcrop in order to gain a full three-dimensional picture. Note that care should be taken to distinguish flow banding in coarse-grained igneous rocks from gneissose banding, developed through metamorphism (see Chapter 12).

Fig. 3.11 Fine flow-banding texture on the surface of obsidian — note conchoidal fracture (right) of this glassy rock. Area of picture is *ca.* 9 × 6 cm.

The origin of feldspar phenocrysts in strongly porphyritic granites is at the centre of an unresolved controversy. The simplest interpretation is that they are early-formed *phenocryst* minerals but some students of granite petrology point to the way they sometimes grow within inclusions, cross the contacts between inclusions and granite, or project into aplite dykes, indicating that they grew at late-magmatic or post-magmatic stages. The possibility of post-magmatic growth of large feldspars has led to the

term *megacryst*, rather than phenocryst, being applied in such cases. The reader interested in this problem will need to couple field evidence, on the relationship of feldspar crystals to granite fabric, with microscopic and geochemical studies in the laboratory.

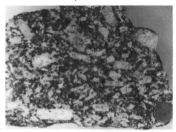

Fig. 3.12 Flow alignment (approximate parallelism) of feldspar phenocrysts in porphyritic granite from Land's End, Cornwall. Area of picture is *ca.* 20 × 12 cm.

A different kind of mineral layering, not simply related to magma flow, is commonly seen in gabbros and also occasionally in diorite and granite intrusions. It takes the form of *alternating* and *repeated* sheet-like layers or thinner (< 2.5 cm) *laminae* richer in light-coloured minerals (Fig. 3.13), and is called *igneous lamination*. Interlocking crystalline igneous textures, with occasional alignment of laminar well-shaped crystals, are preserved throughout this igneous lamination, which is thought to owe its origin to a combination of two processes. The first process involves the preferential accumulation of certain types of crystals, possibly because of density settling rather than their being carried along by magma flow (as in flow banding). This process of large-scale *fractional crystallization* may result in an apparently stratified magma chamber in which there is a complete gradation from the most mafic rocks (i.e. mainly ferromagnesian minerals) at the base to the most felsic rocks (i.e. mainly feldspars with relatively few ferromagnesian minerals) at the top – see Chapter 10 for examples. An igneous rock with a laminar framework of touching mineral grains that were concentrated through fractional crystallization of the parental magmas is a *crystal cumulate* (for detailed terminology, see Irvine (1982) in the further reading list).

Fig. 3.13 Igneous lamination in felsic (lighter) and mafic (darker) gabbro from the Cuillin complex, Isle of Skye, Scotland. Length of hammer is 35 cm.

The second process involved in the origin of igneous lamination, which on a local scale (centimetres to metres) may operate in conjunction with large-scale chemical and mineralogical

stratification in a large magma chamber, is that whereby layers (> 2.5 cm) and laminae (< 2.5 cm) develop by fractionation *in situ* rather than by large scale settling. For example, the fine laminae in Fig. 3.13 probably represent local crystallization of first more-mafic, and then more-felsic minerals in close association. Similar laminae might be forming elsewhere, at the same time, in different parts of a large magma chamber. The field notes describing outcrops of this type should include estimates of the mineral compositions of the most-felsic and most-mafic laminae, since both contain varying proportions of essentially the same minerals, a note on laminae widths, and measurements of their orientations, since these may often be almost parallel to the base of the magma chamber. Later, an overview of the degree of major chemical and mineralogical stratification across a region represented by a layered intrusion may be formed by combining field evidence from many outcrops.

The process of crystallization in large igneous intrusions results in an extended period where crystal–liquid mixtures are present and so may provide a wide range of structures normally associated with sedimentary rocks. For example, Fig. 3.14 shows a series of alternating coarse- and fine-grained gabbros (near the bottom of the picture) overlain by layers of much finer grained jointed gabbro (top left) — the entire sequence has been cut by several generations of aplite and fine-grained gabbro dykes. Occasionally, where layer boundaries are sharp and there is a strong inverted density gradient (e.g. peridotite overlying leucogabbro), plastic deformation may occur before the rock becomes hardened as one layer penetrates the

23

other, resulting in *flame structures* and *load casts* (Fig. 3.15) which look very similar to those of sedimentary formations (see Tucker, 1982 for an explanation of these terms). Similarly, channel structures and block slumping structures may be recognized; other examples of igneous banding, lamination and structures are discussed in later sections, particularly in Chapter 10.

3.4 Summary of field observations

This chapter contains most of the information needed to describe indi-vidual outcrops of igneous rocks in the field. It also contains some preliminary ideas of how you should start to inter-pret your field observations in terms of the history and mode of occurrence of an igneous body. Chapter 4 comple-ments this approach by showing how igneous rock hand specimens should be described and named in the field. Finally, Table 3.1 should be used in conjunction with Table 4.9 as a check-list of observations to be made at each outcrop for easy reference in the field.

Fig. 3.14 Irregularly banded coarse and fine gabbros giving way upwards (and to the left) to fine-grained gabbro and cut by several generations of aplite and gabbro dykes — notably trending from mid-right to top left of the picture. Notice that major joints run at right angles to the strike of the gabbro units and that only the fine gabbros are cut by veins. Cuillin complex, Isle of Skye, Scotland. Length of hammer is 35 cm.

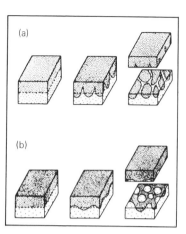

Fig. 3.15 Schematic illustration of the development of (a) flame structures and (b) load casts in laminated igneous rocks subject to plastic deformation caused by reversed density gradients prior to final consolidation (i.e. dark stippled material is heavier than light stippled material). In (a) the light material has lower viscosity than in (b), hence the relatively narrow flame structures and broader load casts, respectively. In both (a) and (b) the sequence from left to right shows the evolution of the structures with time.

Table 3.1 Checklist for describing igneous rock outcrops (to be used in conjunction with Table 4.9, p. 53 for hand specimen desriptions)[1]

1 Record field location, allocate outcrop number and note general form of the outcrop and any distinctive features visible from a distance, making field sketches and taking photographs where necessary.

2 Note orientation and spacing of prominent joint sets and determine whether there is any evidence of faulting.

3 If the outcrop shows veins, aplites, pegmatites, dykes or simple contacts, determine the nature of the different rock-types, especially near contacts; measure the orientation(s) of linear and planar features and deduce the sequence of events from cross-cutting relationships.

4 If the outcrop shows flow banding or igneous lamination, determine the nature and orientation of the banding and, in the case of igneous lamination, examine sharp boundaries between individual units for evidence of 'pseudo-sedimentary' structures.

5 Make detailed notes summarizing the mineralogy and naming provisionally the major rock-types present (cf. Table 4.9).

[1]See Fig. 4.12 for an example of a field description following this scheme: p. 49.

4

Hand specimens and their interpretation

4.1 Collecting field specimens

If an outcrop is believed to contain igneous rocks it will be necessary to make a detailed examination of the rock-type(s) with a view to visiting other outcrops, mapping the shape and form of the igneous body and considering its mode of occurrence. The diagnostic features of igneous rocks are often less easily recognised from weathered than from freshly broken surfaces, though the opposite is sometimes true for coarse-grained mafic rocks. Bear in mind, however, that the effects of weathering may penetrate below the surface to a depth varying from a few millimetres in fine-grained rocks to as much as several metres in the case of coarse-grained or heavily-fissured material and in the case of poorly consolidated pyroclastic rocks. (These depths may be increased by at least a factor of ten in areas which have, or are presently experiencing, wet, tropical climates; Fig. 4.1). Whereas successful field identification may require only small rock chips (a few centimetres in average dimensions) obtained without penetrating the entire weathered skin, samples being collected for microscope and/or geochemical work will need to be much larger (10–20 cm) and as fresh as

possible (see Section 2.4). Unweathered material may have been subject, whilst at depth, to high-temperature, post magmatic hydrothermal alteration which can change the bulk chemical and mineralogical properties of entire igneous bodies and make them more susceptible to surface weathering. The effects of this alteration are recognized most easily through the presence of secondary minerals in hand specimens (Section 4.4).

To what extent can igneous rocks be named in the field? The business of assigning field names to fresh samples may be surprisingly straightforward, because igneous rocks vary widely in their physical characteristics. Even when making a detailed study of a small region in which, for example, successive lava flows may appear remarkably similar at first sight, it is possible with practice and by adopting a systematic approach to develop a keen eye for small changes in the appearance of your samples. The main features of igneous rock samples that should be noted in the field as an aid to identification are:

1 *Colour*, as a preliminary guide to the approximate chemical and mineral composition of the sample.

2 *Texture, grain size* and *fabric* which are related to the rate and sequence of crystal growth.

3 *Mineralogy* — usually between two and four main minerals occur in an igneous rock. For coarse-grained rocks, the number and nature of the main minerals present, their relative abundances and inter-relationships provide a clear guide to rock identification. Many fine-grained rocks also contain some large crystals that may be useful for diagnostic purposes.

In many cases of fine-grained rock identification, it is difficult to make full use of features 1–3 and to record in the field a description that is more detailed than say, 'fine-grained grey lava'. In addition, but not part of the rock identification, there are

4 *Inclusions* of 'foreign' material. These may be found in samples and should be described separately, as they are important when assessing the mode of occurrence and origin of the igneous rock body.

Fig. 4.1 Surface of a deep weathered layer in basalt lava flows at Deserta Grande, Madeira Islands. The layers of basalt are alternately hydrated by rainfall and dessicated by strong solar heating, a process which cracks the rocks and allows greater penetration of water during each successive period of rain. This often results in spherical hydration shells surrounding a core of relatively fresh rock as seen here; the material of the shells has been chemically leached and now comprises soft hydroxide and oxide minerals that are easily worn away. Such spheroidal weathering should not be confused with pillow structure in lava (cf. Fig. 6.5)

The nomenclature for igneous rocks developed in this field guide covers the first three of these criteria, which are used to provide simple field names for rock samples. Several more detailed, but basically similar, classfication schemes are in common use for naming igneous rocks once their accurate chemical and/or mineral compositions are known from *laboratory work*. Some of these schemes use the proportions of the minerals observed under the microscope (their *modal* proportions) whereas others derive, by calculation from the chemical composition of the rock, the theoretical proportions of a series of standard minerals with fixed chemical compositions (the *normative* minerals) and use these as a basis for classification. Although both schemes may be used for intermediate to coarse-grained igneous rocks, glassy and fine-grained rocks are *always* classified accurately in the laboratory on the basis of chemical composition. Often there is little difference between the modal and normative mineral compositions of a rock, but it is recommended that, if your studies extend into the laboratory, you should revise your nomenclature using the system proposed by the IUGS (International Union of Geological Sciences) Subcommission on the Systematics of Igneous Rocks (see Appendix I).

4.2 Colour and composition

Naming of most field specimens requires a working knowledge of the natural variations in their silicate mineral compositions (Section 4.5). Individual silicate minerals have chemical compositions that vary within predictable limits (Table 4.1) and it follows that the mineral and

chemical compositions of rocks are closely related. This relationship is also expressed in the most obvious property of a field specimen — its *colour* — which is reflected in the colours of its constituent minerals.

The following terms may be used to describe the characteristics of individual minerals and the major minerals present within the mode or norm of an igneous rock.

(a) Terms applied to minerals present in the *mode* of an igneous rock and to rocks having one or more of these minerals as major components of the mode:

felsic = *f*eldspar + *le*nad + *si*lica (e.g. feldspars, feldspathoids, quartz) (lenad is a term based on the feldspathoid minerals, e.g. *leu*cite + *ne*pheline)

mafic = *ma*gnesium + ferric (e.g. olivine, pyroxenes).

Mafic *minerals* may also be termed *ferromagnesian* minerals (*ferric* + *magnesium*).

(b) Terms applied to minerals present in the *norm* of an igneous rock and to rocks having one or more of these minerals as major components of the norm:

salic = *s*ilicon + *al*uminium (e.g. quartz, feldspars, feldspathoids)

femic = *f*erric + *m*agnesium (e.g. olivine, pyroxenes)

Although most of your observations will relate to fresh surfaces it is usually worth noting the colour of the weathered skin of an igneous rock, particularly if it is fine-grained. Since the chemical process of weathering tends to concentrate medium- to dark-brown insoluble residues in rocks that contain a high proportion of mafic minerals (Table 4.1), or pale-brown to white residues if most of the minerals are felsic, the colour of the weathered surface provides a first clue to the composition of the rock. Fresh surfaces of most igneous rocks range in colour from dark- to light-grey, greenish-grey or brown, though many of the lighter-coloured rock types also have purple to pink varieties. In simple terms, this dark to light colour range results from the different combinations of dark and light minerals that crystallize from magmas of different compositions (further details in Section 4.5).

After noting the colour of the weathered surface, the next step in classifying an igneous field specimen is to decide where the fresh surface lies within the natural colour range and, if possible, to record its *colour index*, based on your estimate of the total percentage of mafic minerals present (see Fig. 4.2), as follows:

leucocratic (0–33%)
mesocratic (34–66%)
melanocratic (67–100%)

Note that the terms *felsic* or *mafic* may also be used to describe the overall colour of rocks that are, respectively, rich in felsic minerals and therefore light-coloured, or rich in mafic

Table 4.1 Typical chemical compositions[1] for some major silicate minerals in igneous rocks (weight per cent). See Table 4.6 for distinguishing features in hand specimens of these and other minerals.

	SiO_2	Al_2O_3	FeO + Fe_2O_3	MgO	CaO	Na_2O	K_2O	H_2O
Felsic minerals								
Quartz	100	—	—	—	—	—	—	—
Orthoclase	65	18	—	—	—	—	17	—
Albite	69	19	—	—	—	12	—	—
Anorthite	43	37	—	—	20	—	—	—
Muscovite	45	38	—	—	—	—	12	5
Nepheline	42	36	—	—	—	22	—	—
Mafic minerals								
Olivine	40	—	15	45	—	—	—	—
Pyroxene (augite)	52	3	10	16	19	—	—	—
Amphibole (hornblende)	42	10	21	12	11	1	1	2
Biotite	40	11	16	18	—	—	11	4

[1]Silicate minerals are made up of atomic frameworks in which different combinations of cation-forming elements are always bonded to oxygen and so it is customary to quote analyses in terms of simple oxides rather than elements.

Note also that in most mineral groups there is compositional diversity caused by the ability of atomic frameworks to hold different cations in the same site (e.g. substitution between Fe and Mg, or between Na and Ca occurs in many mineral groups).

minerals and therefore dark in colour; these terms are synonymous with leucocratic and melanocratic, respectively.

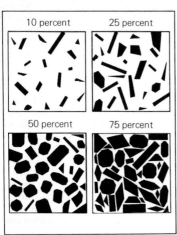

Fig. 4.2 Comparison diagram for estimating the mafic mineral content of two-dimensional igneous rock surfaces. It is well worthwhile using this diagram in the field as there is a common tendency, because of the greater visual impact of dark minerals, for colour index to be overestimated in field reports.

The accuracy with which the colour index can be estimated will depend on the grain size of the specimen. For example, close inspection under a handlens of the fresh surface of a coarse-grained rock should give an estimate within 10% of the true value (e.g. Fig. 4.3), but for fine-grained rocks, the colour index becomes more difficult to determine and is obviously inapplicable to glassy rocks (Section 4.3). In cases where individual crystals cannot be distinguished with a hand-

lens, it is better to record an overall impression of the specimen's colour, thus avoiding any misleading conclusions that could arise later from a poorly estimated numerical value of colour index.

When working in an area of fairly similar coarse-grained rocks, such as a granite intrusion where the colour index might vary from, say, 5% to 25%, then it is permitted field practice to use the prefixes *leuco-*, *meso-* and *mela-* in conjunction with the word 'granite' to describe, respectively, samples with mafic mineral contents that are below average, average and above average for the whole intrusion.

Fig. 4.3 Three samples of equigranular (equal grain sizes) coarse-grained igneous rocks with colour indices of 50, 30 and 10. The top specimen is melanocratic, that on the lower left is on the boundary between mesocratic and leucocratic, and that on the lower right is leucocratic.

Because the colour of an igneous rock reflects the minerals it contains and hence its chemical composition, it follows that the colour index conveys some information about the chemical composition of the rock. It will be obvious from Table 4.1 that the most important chemical constituent of most igneous rocks, expressed in the conventional oxide form, is silica — SiO_2. Almost all igneous rocks contain between 40 and 75% SiO_2 by weight; the higher the silica content, the more felsic minerals the rock must contain and the lower will be its colour index — and vice versa for its content of mafic minerals.

There are obvious similarities between this colour index classification and the much used geochemical classification of igneous rocks into *acid, intermediate* and *basic* categories, using their silica contents. This classification dates from the period when silicate minerals were classified, erroneously, as salts of hypothetical silicic acids. Although the terms acid, intermediate, basic and ultrabasic are defined strictly in terms of the abundance of SiO_2 in igneous rock chemical analyses (Table 4.2), they are also applied loosely in the field; for example, most felsic/leucocratic rocks may be described as acid. Table 4.2 provides a ready comparison of the different terms used in describing the colour of an igneous rock; in general, it is best to use those on the right for field descriptions.

Although the implications of colour for an assessment of rock composition have been described in some detail, clearly there are many other clues to be considered before assigning field names to igneous rock samples. Therefore, do not jump to hasty conclusions but, after noting the colour of the weathered and fresh surfaces, consider next the other features of field specimens — their texture and mineral content.

Table 4.2

Geochemical term	Definition wt % SiO_2	Approximate range of colour index[1]	Possible field descriptions
Acid	> 65	5–25	Leucocratic or felsic
Intermediate	52–65	25–55	Mesocratic
Basic	45–52	55–85	Melanocratic or mafic
Ultrabasic	< 45	85–100	

[1]Note that there is a general correlation, in which felsic/salic rocks are leucocratic, with low colour index and of acid chemical composition, whereas mafic/femic rocks are melanocratic, with high colour index and of basic/ultrabasic chemical composition. However, there are exceptions to this generalization. For example, rocks composed of Ca-rich plagioclase, termed *anorthosites*, are felsic/salic leucocratic rocks with low colour index but of *basic* or intermediate chemical composition. Similarly, rocks composed of Na-, Ca-, Mg- carbonates, termed *carbonatites*, are leucocratic rocks with low colour index and of *ultrabasic* chemical composition.

4.3 Texture, grain size and fabric

Collectively, the relationships between the constituent crystals and glass (where present) in an igneous rock are known as *texture*. When describing the texture of a field sample, examine:

1 The *grain-size* of the rock which can reflect the rate at which it crystallized.

2 The *fabric*, or geometrical characteristics and arrangement of the crystals including, where possible, observations of the number of minerals present and the characteristic shapes, or *habits*, of their crystals.

3 The overall *homogeneity* of the specimen (i.e. whether it is uniform and equigranular, or contains mineral segregations, banding, and irregular inclusions).

These features all provide clues to the physical conditions under which the magma crystallized; thus, whereas the colour of a specimen is generally related to its chemical composition (e.g. granites are usually lighter in colour than gabbros), the texture reflects its history and mode of occurrence (further details in Chapter 5). For example, rapid chilling leads to the formation of a glass, whereas the nucleation of crystals at many sites in a magma erupted at the surface and still subject to fast cooling, such as in lava flow, tends to result in minute grains that usually cannot be identified in the field. On the other hand, slow cooling, as in intrusions emplaced below the surface, is accompanied by crystallization at relatively few nucleation sites and results in larger, easily observed crystals.

Grain size is usually the most obvious textural characteristic, and after examining the specimen with the aid of a handlens, it should be assigned to one of the categories in Table 4.3.

The terms in Table 4.3 are easily applied to rocks that are more or less *equigranular* (equal grain sizes), but many specimens are porphyritic (Fig. 4.4), having a population of early-formed, large *phenocrysts* in a finer grained *groundmass* or *matrix*. In such cases, the actual size of both populations of crystals should be recorded. Note that another short-hand way of describing porphyritic texture, once the phenocryst mineral has been identified, is to use the hyphenated adjective '-phyric' along with the mineral name: thus Fig. 4.4a shows a *feldspar-phyric* granite. But it may prove impossible to continue the description of non-porphyritic (or *aphyric*) specimens in this category without recourse to laboratory techniques. Field names such as 'fine-grained felsic or mafic rock' are recommended on the basis

Table 4.3

Fine-grained*	Few crystal boundaries distinguishable in the field or with the aid of a handlens; mean grain size below 1 mm. If the rock is glassy, the term *hyaline* may be used.
Medium-grained**	Most crystal boundaries distinguishable with the aid of a handlens; mean grain size 1–5 mm.
Coarse-grained**	Virtually all crystal boundaries distinguishable with the naked eye; mean grain size greater than 5 mm.

*Fine-grained and hyaline rocks may be termed *aphanitic* in texture.
**Medium- and coarse-grained rocks may be termed *phaneritic* in texture.

(a)

(b)

Fig. 4.4 Examples of porphyritic texture: (a) Rectangular phenocrysts of feldspar (note twinning in crystal near the bottom), about 2 cm in longest dimension, in granite. The coarse-grained groundmass contains more feldspar together with grey, glassy quartz grains and black plates of biotite mica. (b) Black, partly lozenge-shaped phenocrysts of amphibole in an otherwise aphanitic sample of grey andesite lava. Area of photograph 9 × 6 cm.

that the former are likely to contain mainly feldspars whereas the latter will contain mainly mafic minerals.

The next stage in describing medium and coarse-grained rocks is to examine their *fabric*, that is, the *shape* of the crystals and their *relationship* to one another. First compare and contrast the characteristics of different types of crystals in the rock. Minerals left to crystallize 'freely' in a magma will form well-shaped crystals which, in the two dimensions of a field sample surface, would show regular, straight-sided cross-sections, often of rectangular shape (but see Table 4.6 for common crystal habits). During the crystallization of most igneous rocks several minerals are competing for the available space and only the first-formed crystals, such as any phenocrysts present, will have good crystal forms. Three sets of terms are used to

Table 4.4

Preferred terms	Synonymous terms	Synonymous terms	Definition
Euhedral	Idiomorphic	Automorphic	Crystal completely bounded by its characteristic faces
Subhedral	Hypidiomorphic	Hypautomorphic	Crystal bounded by only some of its characteristic faces
Anhedral	Allotriomorphic	Xenomorphic	Crystal lacks any of its characteristic faces

describe the quality of development of crystal faces, the most commonly used set being that in the left-hand column of Table 4.4 (see also Fig. 4.5).

In ideal circumstances, where the grain shapes of a coarse-grained rock are extremely clear, it may be possible to classify the grain shape fabric of the entire rock. Depending on the general shape of the crystals, three textures can be distinguished within equigranular rocks, as shown in Table 4.5.

The boundaries between the categories defined in Table 4.5 are not sharply defined and consequently the terms are applied very subjectively. Furthermore, a rock may not fit neatly into a single category: thus, one in which ~ 50% of the crystals are euhedral and ~ 50% anhedral might best be described as having a mixed euhedral and anhedral granular texture.

After describing the individual crystals and their inter-relationships, the remaining textural characteristics to notice concern the overall *homogeneity* of the rock. At the scale of a hand

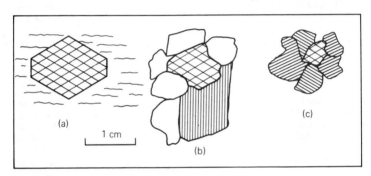

Fig. 4.5 Sketches of (a) euhedral, (b) subhedral and (c) anhedral amphibole crystals (cleavage traces at 60° and 120°, cf. Table 4.6).

Table 4.5

Term	Synonyms	Definition
Euhedral granular	Panidomorphic granular	Bulk of the crystals are euhedral and of uniform size
Subhedral granular	Hypidiomorphic granular	Bulk of the crystals are subhedral and of uniform size
Anhedral granular	Allotriomorphic granular (granitic and granitoid are textures that apply to siliceous rocks only)	Bulk of the crystals are anhedral and of uniform size

specimen, most igneous rocks are fairly homogeneous but occasionally *flow banding* or *mineral lamination* are encountered (Section 3.3). Flow banding in medium and coarse-grained rocks may be expressed through the visible alignment of elongate crystals (Fig. 3.12) or, in fine-grained rocks, by a fine-ribbing or colour banding which is often most noticeable on weathered surfaces (Fig. 3.11). As flow structures are developed parallel to the direction of movement, measurement of their orientation will act as a useful aid in mapping lava sequences. Mineral lamination, expressed as alternating lighter and darker layers in coarse-grained rocks, may originate through fractionation processes (as described in Section 3.3 and Fig. 3.13) or, perhaps, through high-grade metamorphism of otherwise homogeneous igneous material (see Chapter 12 for details). Subsequent interpretation of field data in all these cases is made a great deal easier if careful notes are made in the field of the colour, texture, orientation and scale of the layering, not only in hand samples, but also in the outcrop as a whole.

Fine-grained lavas often contain randomly scattered spherical or subspherical cavities, usually measuring a few millimetres in cross-section and known as *vesicles*. *Vesicular* texture and the associated phenomenon where the cavities are filled with secondary minerals — *amygdaloidal* texture — are discussed in Chapter 6 (Fig. 6.3).

The textures which have been introduced in this section are those frequently encountered in igneous field studies; the list will serve as a general guide in the field but will not cover all occurrences. Even if you do not recognize the textural features of a particular hand specimen, remember that the golden rule is to *make accurate observations backed up by field sketches as appropriate*. More examples of these and other less frequently observed textures at the scale of hand specimens appear later in this field guide.

4.4 Mineral identification

The most informative, yet also the most difficult, part of a field description of igneous rocks is the identification of the constituent minerals. Igneous rocks consist of abundant or *major* minerals, together with minor minerals that are much less abundant. Major and minor minerals may be termed *essential* and *accessory* minerals. Essential minerals are those that are necessary to the naming of the rock. They are usually major minerals, for example, quartz in a granite, but are sometimes present in minor amounts, for example olivine, in an olivine basalt. Accessory minerals are those which are present in such small amounts that they are disregarded in the definition of the rock; for example, Fe-Ti oxide in a gabbro, or biotite in a granite.

Even though most rocks contain only two or three essential minerals, together with a few accessory minerals, *it is rarely easy to identify all the essential minerals and often impossible to name the accessory minerals*. The problem is most acute for fine-grained mafic rocks in which many of the mafic minerals have a similar appearance. Even in coarse-grained leucocratic rocks, it is difficult to distinguish among the different minerals, particularly between plagioclase and alkali feldspar. However, as you gain in experience of the properties of minerals and of the types that commonly occur together, the more confident your naming of rocks will become. It is worth remembering that most igneous rocks represent an assemblage of minerals that crystallized almost in chemical equilibrium with one another from a magma. So, for example, the mafic minerals olivine and pyroxene *or* biotite and amphibole frequently occur together but other combinations, such as olivine and biotite, are much less common. The kinds of combinations that occur most frequently in igneous rocks are summarized in the mineral-rock classification diagrams on the following pages; note, in particular, Fig. 4.7. *For field purposes, it is seldom necessary to spend time identifying more than one or two of the main minerals in a rock specimen; this can be achieved with much greater success with the microscope.*

In the course of observing the colour and texture of a rock you will have formed an impression of the approximate number and some of the properties of the individual minerals. The next stage is to identify some of those minerals with the aid of Table 4.6, which summarizes mineral properties under five major headings: colour, cleavage, lustre, habit and hardness. Generally, it is best to start with the light-coloured felsic minerals, as these are often most easily identified. A few points about the use of Table 4.6:

(i) *Colour* is an obvious property of a mineral which, like that of a rock (Section 4.2), reflects its overall chemical composition. Generally, alkali/alkaline earth silicates (Na, K, Ca silicates) are light-coloured whereas silicate minerals rich in transition elements, particularly iron, are dark-coloured.

(ii) *Cleavage* is the tendency of minerals to split along well-defined planes that are related to weaknesses in their atomic structures. Cleavage surfaces are flat planes that reflect light evenly and are often responsible for lustre (see below); for example, freshly-broken feldspar crystals may show slightly jagged edges which, under a handlens, are seen to be a composite of many flat cleavage surfaces intersecting at or near 90° (Fig. 4.6). The perfect single cleavage of mica, which forms the brightly reflecting surfaces, often allows small flakes to be dislodged with a penknife or finger nail. In contrast, cleavage in pyroxenes and amphiboles is seen only occasionally as fine striations within the grains, whereas the characteristic cross-sections of euhedral crystals (8-sided and 6-sided lozenge shapes, respectively, for pyroxenes and amphiboles, Table 4.6) are much more diagnostic.

(iii) *Lustre* describes the reflective properties of a mineral and is assessed by turning the specimen until the mineral surfaces, particularly any cleavage surfaces, are caught by the light. Lustre varies from *dull* (non-reflecting), through *resinous* and *silky* to *bright*, which may be metallic, glassy or vitreous (which literally also means glassy but here is defined slightly differently) depending on the mineral concerned. A brightly coloured silicate mineral which is opaque to the transmission of light may have a *vitreous* lustre, like glazed pottery, such as clean crystals of amphibole and pyroxene. On the other hand, a similar translucent (weakly transparent) or transparent mineral may appear *glassy* (e.g. quartz and fresh olivine).

(iv) *Habit* refers to the characteristic morphology of euhedral crystals. Of course, most crystals in igneous rocks are not euhedral and, given the random mixture of cross-sections through three-dimensional mineral grains represented in a hand-specimen, many grains may need to be examined before a characteristic shape is found (e.g. pyroxene and amphibole habits are distinctive only when viewed close to one particular planar cross-section). Some minerals have quite well-defined characteristic shapes in most euhedral cross-sections, e.g. lath-shaped feldspars which are long and thin, but rectangular, dodecahedral or trapezohedral garnets, and acicular (i.e. needle-like) tourmalines. Others, such as quartz and olivine, are less commonly euhedral. *Twinning* results from crystal growth in different orientations on either side of a common atomic structural plane. It is very occasionally visible in large crystals of feldspar, where it provides the only practical way of distinguishing alkali feldspar (simple twins only, Fig. 4.4a)

from plagioclase (simple and multiple twins) in hand specimen (see Table 4.6). Twinning is best observed by rotating cleavage or crystal faces in sunlight; individual twins are seen to have slightly different angles for optimum reflection and, at the hand specimen scale, are commonly seen only in alkali feldspars.

(v) *Hardness*, a property related to the strength and uniformity of atomic bonding within a crystal, is described using a simple but non-linear scale of increasing hardness from 1 to 10 known as *Mohs' scale*. If the mineral grains are large and fresh enough (weathered grains are often softer than the original), then a series of scratching tests using a finger nail (hardness *ca.* 2), a copper coin (hardness *ca.* 3.5) and a steel penknife blade or geological hammer (hardness *ca.* 5.5) may prove diagnostic. For example, mica or calcite will scratch easily with a coin or penknife; feldspars may scratch a penknife blade with difficulty, but quartz will always scratch it easily.

(a)   (b)

Fig. 4.6 Cleavage in (a) feldspar which has two good cleavages at about 90° and (b) mica which has one excellent cleavage, here parallel to the plane of the page.

Table 4.6 Mineral properties in igneous rock hand specimens

Mineral	Typical chemical formula	Colour	Cleavage	Lustre	Habit	Hardness
Felsic minerals						
Quartz	SiO_2	Colourless to pale grey when surrounded by dark minerals; transparent	None; irregular, or curved fracture surfaces	Glassy, shiny	Rare trigonal pyramids but usually irregular, anhedral	7
Alkali feldspar	$(K,Na)AlSi_3O_8$	White or pink, sometimes orange or yellow	2 sets at 90°, poorly visible	Usually dull, sometimes silky or vitreous	Tabular crystals; shiny cleavage surfaces may show simple twins. Elongate rectangular 'laths', lamellae, or irregular masses of plagioclase may be noted: perthite	6
Plagioclase feldspar	$NaAlSi_3O_8$ to $CaAl_2Si_2O_8$	White or green, rarely pink or black	2 sets almost at 90°, poorly visible	Usually dull, sometimes silky or vitreous	Lath-shaped crystals; shiny cleavage surfaces may show multiple, parallel twins	6–6.5
Nepheline	$NaAlSiO_4$	White to pale grey	2 poor cleavages, 1 occasionally distinct	Greasy, vitreous	Usually occurs in microcrystalline groundmass; occasional aggregates of crystals	5.5–6
Muscovite (mica)	$KAl_2(AlSi_3O_{10})(OH)_2$	Colourless to pale brown or green	1 excellent cleavage, cleaves into thin flexible sheets	Shiny, silver and pearly	Tabular crystals sometimes 6-sided, especially in pegmatites	2–2.5

Mafic minerals

Mineral	Formula	Colour	Cleavage	Lustre	Crystal form	Hardness
Olivine	$(Mg,Fe)_2SiO_4$	Olive green, yellow-green, sometimes brown	Very poor, usually fractures	Glassy when fresh, vitreous when altered	Usually rounded anhedral crystals, occasionally equidimensional tabular forms	6–7
Pyroxene	(i) $(Mg,Fe,Ca)_2Si_2O_6$ (augite, etc.) (ii) $NaFeSi_2O_6$ (aegirine)	Black to dark green or brown Yellowish-green	2 good sets meeting at nearly 90°	Vitreous when fresh, dull when altered	4- or 8-sided prismatic crystals occasionally showing cleavage or Aegirine more acicular	6
Amphibole	(i) $Ca_2(Mg,Fe)_5Si_8O_{22}(OH)_2$ (e.g. tremolite) (ii) $Na_2Fe_3^{2+}Fe_2^{3+}Si_8O_{22}(OH)_2$ (riebeckite)	Black to brownish black or dark green Dark blue	2 good sets meeting at 120°	Vitreous when fresh, dull when altered	Prismatic or lozenge-shaped crystals often showing cleavage or Riebeckite more acicular	5–6
Biotite (mica)	$K(Mg,Fe)_3(AlSi_3O_{10})(OH)_2$	Black to dark brown or green	1 excellent cleavage; cleaves into thin flexible sheets	Very shiny	Thin tabular crystals, occasionally 6-sided, especially in ignimbrites and acid lavas	2.5–3
Tourmaline	$Na(Mg,Fe)_3Al_6B_3Si_6O_{22}(OH,F)_4$	Black, but varieties may be blue, red or green	Very poor	Vitreous shiny	Long thin prismatic needle-shaped crystals, sometimes longitudinally striated and often in clusters; occasionally striated curved surfaces	7
Frequent accessory minerals						
Apatite	$Ca_5(PO_4)_3(OH)$	Pale green to yellow green	Very poor	Vitreous	Often euhedral, sub-hexagonal crystals; sometimes fibrous	5
Sphene	$CaTiSiO_4(OH)_2$	Colourless to yellow, green or brown	1 good cleavage	Vitreous	Characteristic euhedral rhombic crystals	5

Table 4.6 Mineral properties in igneous rock hand specimens (continued)

Mineral	Typical chemical formula	Colour	Cleavage	Lustre	Habit	Hardness
Frequent Accessory Minerals (contd.)						
Garnet	$(Mg,Fe)_3Al_2Si_3O_{12}$ (also Ca, Mn)	Red, brown or yellow	Poor	Usually resinous or dull, good crystals may be glassy	Equidimensional crystals often showing faces typical of cubic system, e.g. dodecahedra and trapezohedra. Common in metagranites.	6–7
Leucite	$KAlSi_2O_6$	White or grey	None	Vitreous or resinous	Often euhedral trapezohedral crystals in alkaline lavas	5.5–6
Hematite	Fe_2O_3	Red to red-brown, sometimes black	None	Dull	Usually fine and powdery, occasionally scaly or fibrous crystals	5.5–6
Magnetite, (a spinel mineral)	Fe_3O_4	Black, brownish-black	Poor	Metallic, dull	Small equidimensional granular crystals, occasional cubes or octahedra	5.5
(Other spinels are $M^{2+}M^{3+}_2O_4$ where M^{2+} is Fe, Mg, Mn, Zn, etc. and M^{3+} is Al, Fe, Cr, etc. e.g. dark brown to black *chromite*, $FeCr_2O_4$, which occurs in some peridotites)						
Ilmenite	$FeTiO_3$	Black, brownish black or grey	None	Metallic or dull	Thin plates or scales usually elongate crystals, sometimes rod-like	5.6
Monazite	$(Ce,La,Th)PO_4$	Pale yellow to dark brown	Moderate single cleavage	Resinous	Thick tabular crystals in granites and gneisses	5–5.5

40

Secondary minerals (See also Table 8.2)

Calcite	$CaCO_3$	White, translucent	3 sets rhombohedral	Vitreous, rarely glassy	Usually granular or fibrous in igneous rocks, common in veins, cavities, etc. NB Reacts with dilute acid	3
Zeolite group	e.g. $(Na_2Ca)(Al_2Si_5O_{10}).$ nH_2O (n varies from 2 to 8)	White, pale yellow or pale green, rarely pink, red or blue	Variable according to mineral type	Usually vitreous or silky	Massive or granular crystals lining cavities, particularly amygdales; radiating fibrous clusters or needles	5–6
Clay group	e.g. $Al_4Si_4O_{10}(OH)_8$ (kaolinite)	White to pale browns and greens	Good, but not visible in hand specimens	Dull	Fine powdery aggregates replacing mainly feldspar in igneous rocks	1
Epidote	$CaFe^{3+}Al_2Si_3O_{12}(OH)$	Pale yellows and apple green, rarely brown or red	1 good cleavage	Vitreous	Variable, often elongated crystals, needles and radiating groups, coarsely crystalline varieties in hydrothermal veins and vesicles	6–7
Chlorite	$(Mg,Al,Fe)_6(Si,Al)_8O_{20}$ $(OH)_4$	Mid-green to dark greenish-yellow	1 good cleavage gives thin sheets	Dull to pearly and 'micaceous'	Usually aggregates of fine crystals, sometimes thin tabular flakes replacing mafic minerals in igneous rocks	2–3
Pyrite	FeS_2	Brassy yellow, occasionally brown or black	Poor	Metallic; iridescent tarnish	Often good cubic crystal faces, occasionally striated. Granular aggregates, particularly along veins in igneous rocks	6–6.5

Table 4.6 contains sufficient information for good examples of the ten most common minerals that occur in coarse-grained igneous rocks to be identified. Also included are eight of the more abundant accessory minerals whose individual abundances rarely exceed 10% by volume of a rock. Whereas apatite, sphene, garnet and leucite sometimes form easily identified euhedral crystals, the three opaque iron-titanium oxide minerals, which often fill the interstices between major silicate minerals, have similar appearances in hand specimens. The list of *secondary* (i.e. post-magmatic) minerals includes (a) those found as amygdale minerals in lavas (e.g. calcite, zeolite); (b) those found in hydrothermal veins cutting igneous rock outcrops (e.g. calcite, chlorite, but also quartz), and (c) those which *replace* primary silicates in rocks that have been subject to pervasive hydrothermal or metasomatic alteration and weathering reactions at submagmatic temperatures (e.g. clay minerals and/or epidote replacing feldspars and mica; chlorite replacing mafic minerals).

4.5 Naming field specimens

When describing a medium- or coarse-grained field specimen you should estimate, as a percentage, the volume of the rock occupied by each identified mineral and, for those you are unable to identify, make a brief note of their properties for later reference. Observations of colour, texture and mineral content can now be combined into a field name for the rock specimen using, initially, Fig. 4.7. Earlier, it was noted that the natural chemical range of most igneous rocks from 40 to 75% SiO_2 is accompanied by a progressive

increase in their content of felsic minerals and a corresponding decrease in their mafic constituents. This information appears on the front face of Fig. 4.7 with the acid, silica-rich rocks on the left and the ultrabasic rocks on the extreme right.

Although this scheme for preliminary rock identification works well in the field for medium- and coarse-grained rocks, it is often possible to follow more closely the precise definitions of the IUGS Subcommission (Appendix 1). This scheme classifies all rocks with a content of mafic minerals (plus Fe-Ti oxides, carbonates of igneous origin and accessory minerals) of less than 90%, primarily using the relative proportions of their different felsic minerals: Q = quartz, A = alkali feldspars, P = plagioclase feldspars, F = feldspathoid minerals (i.e. nepheline and leucite in Table 4.6, though there are other feldspathoids; these minerals have similar chemistry to feldspars but with less SiO_2 and are often collectively known by the abbreviation *foid* minerals). Figure 4.8 is a simplified version of the QAPF diagram which, with experience, can be used in the field quite successfully, especially in conjunction with Figs 4.9 to 4.11.

Consider first those rocks where *quartz is typically 20–60% of the felsic minerals*: these can be named accurately as *alkali granite, granite, granodiorite* or *tonalite*, if the relative proportions of alkali feldspar and plagioclase are estimated (cf. Fig. 4.9). Where this is not possible, any sample with 20–60% quartz may, provisionally, be termed '*granitoid*'. Figure 4.9 illustrates the usual range of mafic minerals present in granitoid rocks and the areas in which the prefixes leuco- and mela- would normally be applied. Similarly, *the feldspars of rocks with lesser* amounts of quartz, or no quartz and small amounts of feldspathoid minerals may allow a distinction to be made between *alkali syenite, syenite* and an abundant group of plagioclase-rich rocks — the diorites, gabbros, and a less common group discussed later, the anorthosites (compare Figs 4.8 and 4.10). Strictly, the distinction between diorite and gabbro lies in the composition of the plagioclase feldspar which must contain more than 50% of the albite (NaAlSi$_3$O$_8$) molecule in diorite and more than 50% of the anorthite (CaAl$_2$Si$_2$O$_8$) molecule in gabbro. As different plagioclase feldspars cannot be identified in the field, the use of a secondary characteristic is recommended — the nature of the common mafic mineral (Fig. 4.7) — to make a

preliminary distinction between *diorite* (with biotite and amphibole) and *gabbro* (usually with pyroxene). It is emphasized once more that, wherever possible, these provisional field names should be *re-examined* in the light of subsequent microscope studies using a more detailed version of the QAPF diagram (Fig. 4.8, Appendix I).

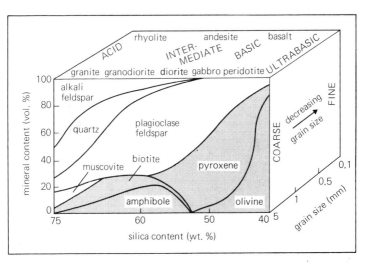

Fig. 4.7 Preliminary classification scheme for igneous rocks using colour index (front face shows approximate proportions of light and dark minerals that occur at different silica contents), common silicate minerals (where recognized in the field) and grain size (decreasing with depth into the diagram).

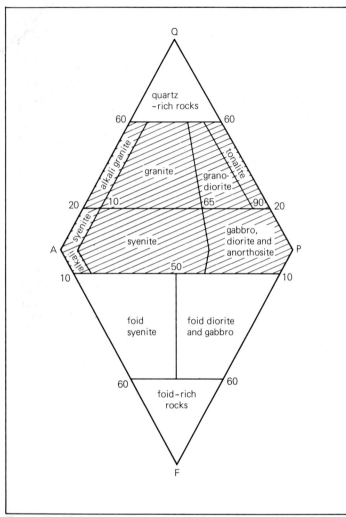

Fig. 4.8 Simplified QAPF diagram for naming medium- and coarse-grained rocks in the field using their salic mineral content (see text and Appendix I for further details). Most igneous rocks fall in the shaded area. Remember that foid is an abbreviation for feldspathoid.

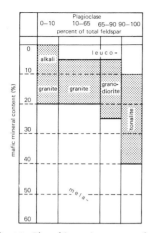

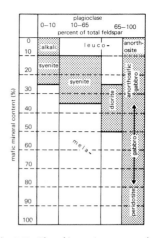

Fig. 4.9 Plot of increasing content of plagioclase feldspar (left to right), as a percentage of total feldspar, against mafic mineral content for granitoid rocks, defined as those with quartz as between 20 and 60% of the salic minerals.

Fig. 4.10 Plot of increasing content of plagioclase feldspar (left to right), as a percentage of total feldspar, against mafic mineral content for common rocks with less than 20% of the salic minerals as quartz. The difference between diorite and gabbro is discussed in the text and the names used here are defined more precisely in later chapters.

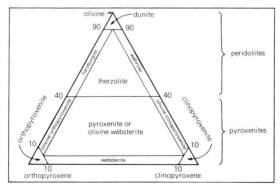

Fig. 4.11 Triangular plot showing proportions of olivine, orthopyroxene and clinopyroxene for classification of ultramafic rocks (rocks with 90% or more of mafic minerals). The commonest ultramafic rocks are lherzolite and harzburgite.

45

Among the rock-types introduced so far are two major acid-intermediate-basic geochemical series, or associations, the occurrence of which tends to be mutually exclusive (hence the separate discussion of them in Chapters 8 and 9).

(i) the abundant granite-granodiorite (tonalite-diorite-gabbro series (cf. Fig. 4.7), often known as the *calc-alkaline* association;

(ii) the less abundant alkali granite-alkali syenite-syenite-foid syenite (usually nepheline syenite) series, known as the *alkaline* association.

These associations have strict geochemical definitions based on chemical ratios, such as $CaO/(Na_2O + K_2O)$ (Cox, Bell and Pankhurst, 1979), but, as will be clear from Figs. 4.8—4.10, the geochemical differences are reflected in the nature of the feldspars and, less obviously, in the mafic mineral content. Rocks of the alkaline series have relatively low abundances of mafic minerals (e.g. alkali syenite and syenite in Fig. 4.10) and, instead, their low silica (basic) end-members extend closer to the feldspathoid apex in Fig. 4.8 — nepheline is a typically abundant mineral. In contrast to the dark-browns and black colours of pyroxenes and amphiboles in diorites and gabbros, the smaller amounts of mafic minerals of alkali rocks are often distinctive, consisting of spectacular green or dark-blue alkali pyroxenes and amphiboles (e.g. aegirine and riebeckite — compare formulae with common pyroxenes and amphiboles in Table 4.6). Thus a secondary characteristic distinguishing alkali granite from calc-alkaline granite, for example, is often the presence of dark-blue amphibole needles in the former and of black prismatic amphiboles and/or

micas in the latter (see also Chapters 8 and 9).

In the low silica, ultrabasic rocks, there are two important groups to consider (cf. Fig. 4.10):

(i) a leucocratic group, the *anorthosites*;

(ii) a melanocratic group, the *peridotites* and *pyroxenites*.

Anorthosites are pale-brown or white monomineralic rocks dominated by plagioclase feldspar; varieties with more than *ca.* 15% of mafic minerals are termed anorthositic gabbros. They are sometimes found in association with alkaline rocks such as syenites and nepheline syenites and sometimes with layered mafic-ultramafic rocks (further details in Chapters 9 and 10). In contrast with anorthosites, rocks that are virtually devoid of feldspar but have abundant mafic minerals (> 90%), and range from dark-green or brown to black in colour, are known collectively as *ultramafic*. These are subdivided further, according to their particular combinations of mafic minerals, into pyroxenites (> 60% pyroxene), peridotites (> 40% olivine) and several other minor categories that are not easily recognized in the field (Fig. 4.11). *Dunite*, an olive-green monomineralic rock with 90–100% olivine, is perhaps the most distinctive, but recognition of the other types depends on the separate identification of orthopyroxene and clinopyroxene which is sometimes difficult in hand specimens. Orthopyroxenes are generally of paler brown or green colours than clinopyroxenes and sometimes weather to a red-brown colour whereas, with a few rare exceptions, clinopyroxenes are very dark brown, green or black. Green orthopyroxene is distinguished from olivine using crystal shape and cleavage (Table 4.6). However, a particularly distinctive clinopyroxene that occurs in some ultramafic rocks in bright green *chrome diopside* (CaMgSi$_2$O$_6$ containing a few per cent of the NaCrSi$_2$O$_6$ molecule).

The igneous rocks considered so far are all composed essentially of silicate minerals. However, a rather rare group of igneous rocks, known as *carbonatites*, contain more than 50% (by volume) of carbonate minerals and occur as both extrusive lavas and intrusive dykes and plugs. These are subdivided further using the major carbonate mineral present into calcite (CaCO$_3$)-carbonatites, known as *sövites*; dolomite (CaMg(CO$_3$)$_2$) carbonatites, known as *beforsites*, which tend to be cream or pale-yellow; and *ferrocarbonatites* where yellow-brown siderite (FeCO$_3$) is the main mineral. Although these igneous rocks are rare, they are associated with alkaline igneous rocks in many areas.

The terminology introduced so far in this section (cf. Figs 4.8–4.11) applies exclusively to coarse-grained rocks. Finer grained examples in which the requisite minerals can be identified in hand specimen should be given the same names with the prior adjectives (Table 4.3, but see also Table 4.8) *'medium-grained'* or *'fine-grained'* (e.g. fine-grained granite). However, in the special case of fine- or medium-grained mafic intrusive rocks, the terms *dolerite* (or *diabase* in the USA) are in common use and are equally as acceptable as fine- or medium-grained gabbro.

Although most medium-grained rocks can be matched in mineral composition with the common rock-types described above, there is one group that occurs in minor intrusions which has distinctive mineralogy and texture:

the *lamprophyres*. These are medium-grained mesocratic or melanocratic (occasionally ultramafic) porphyritic rocks with phenocrysts of biotite and/or amphibole in a groundmass containing feldspars and/or feldspathoid minerals. They may contain hydrothermal calcite and zeolites, and the primary igneous minerals are frequently extensively altered to these minerals. They are difficult to classify further due to the fine grain size and alteration of the groundmass: a simplified scheme is given in Table 4.7.

Fine-grained rocks have an individual set of names, although for each coarse-grained rock type there is a fine-grained equivalent (Table 4.8). The distinctive mineralogy of different porphyritic rocks in this category, together with their colour, often provides enough information for them to be named in the field; it is customary to use their phenocryst minerals as descriptive adjectives (see Table 4.8). Not all the phenocryst minerals listed in Table 4.8 need occur in each of the respective rock-types; indeed, most porphyritic rocks are named on the basis of the phenocrysts present, e.g. plagioclase, pyroxene-phyric andesite, olivine-phyric basalt. In cases of doubt, and particularly for aphyric specimens, the terms fine-grained felsic or mafic rock are recommended. The nomenclature of volcanic rocks is discussed in more detail in Chapter 6. Finally, *Table 4.9 provides a checklist for easy reference when describing and naming igneous rocks in the field; Fig. 4.12 gives an example of such a description in conjunction with observations of the outcrop from which the sample was taken (the technique used is described in Chapter 3 and summarized in Table 3.1).*

Table 4.7 Classification of lamprophyres

Felsic minerals		Mafic minerals		
Feldspar	Foid	Biotite, pyroxene ± olivine	Amphibole (hornblende), pyroxene ± olivine	Amphibole (barkevite, kaersutite[1]), pyroxene, olivine, biotite
Alkali feldspar > plagioclase	—	Minette[2]	Vogesite[2]	—
Plagioclase > alkali feldspar		Kersantite[2]	Spessartite[2]	
Alkali feldspar > plagioclase	Feldspar > foid	—	—	Sannaite[3]
Plagioclase > alkali feldspar	Feldspar > foid	—	—	Camptonite[3]
No feldspar	Glass or foid	—	—	Monchiquite[3]

[1]Kaersutite is a Ti-, Mg-rich amphibole common in alkaline volcanic rocks.
[2]These are termed calc-alkaline lamprophyres. [3]These are termed alkaline lamprophyres.

Table 4.8 Fine-grained rock-types, their medium- and coarse-grained equivalents, and possible phenocryst minerals (see also Table 6.1, p.67)

Fine-grained rocks	Medium- and coarse-grained equivalents	Common phenocryst minerals
Felsic (leucocratic)		
Rhyolite	Granite	Quartz, alkali feldspar
Dacite	Granodiorite	Quartz, plagioclase feldspar
Trachyte	Syenite	Alkali feldspar, occasional mafic minerals
Phonolite	Nepheline Syenite	Alkali feldspar, nepheline or leucite, occasional green alkali pyroxene
Mafic (meso- to melanocratic)		
Andesite	Diorite	Plagioclase, pyroxene or amphibole, occasionally quartz or biotite
Basalt	Gabbro	Plagioclase, pyroxene, olivine

ARDNAMURCHAN 5-11-83

Roadcut 800m SW of Achnaha village, east side of road Ge u67680; outcrop no 83-11-126

Discontinuous outcrop 30 x 2m high (max), of massive apparently homogeneous dark grey rock with flat-lying joints cut by less homogeneous medium brown rock, ca. 15m wide dyke (?) dipping ca. 70° to S (away from main centre already mapped to N) — hence possible ring dyke. Dyke rock shows regularly dipping joints and some flat-lying joints.

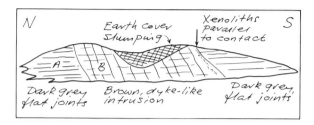

Joints — flat-lying, spaced ca. 20-30 cm, density increases towards top of outcrop. Dipping joints in dyke about every metre but relatively diffuse in centre of outcrop.

Contacts are sharp but xenoliths of host occur within 10-20cm marginal zone of dyke on S margin only. Average size of xenoliths ca. 5cm, apparently unmelted & quite angular. Dip of contact 275/72° (S. margin), 268/74° (N. margin).

Chilled margins: grain size within dyke is uniformly medium (ca. 1-2 mm average) except within 10 cm of margin where it becomes finer but not aphanitic.

SAMPLES A = 8311126A   B = 8311126B

A. Coarse-grained uniformly equigranular rock with little difference between fresh & weathered surfaces. ca. 50% euhedral light grey-green feldspar, resinous, visibly twinned crystals, ca. 20% irregular anhedral olivine (dark green), ca. 30% black pyroxene; occasional silver-grey metallic grains (ilmenite): olivine gabbro. Slight variations within outcrop to 30% feldspar, 70% mafics but no obvious banding.

B. Medium to fine-grained porphyritic rock, colour index ca. 30 — phenocrysts of lath-shaped plagioclase feldspar comprise 20% of rock; they average 2mm long and show parallelism with dip of dyke. Possible quartz + feldspar comprise 50% of ground mass, remainder is dark brown vitreous mineral (black on fresh surface) — pyroxene? — Dyke material may be feldspar-phyric quartz dolerite. Varies to lighter brown, finer grained material at margins.

Fig. 4.12   Example of an entry from a field note-book showing details of an outcrop in Ardnamurchan, Scotland, using the schemes given in Tables 3.1 (p.25) and 4.9 (p.53).

4.6 Inclusions in field specimens

During the ascent of magma bodies to their level of emplacement or extrusion, fragments of solid rock material from the surrounding region or from the walls and roof of the magma chamber may become incorporated and remain as foreign material in the solidified igneous rock. These foreign bodies are known as *xenoliths*, or sometimes, as *enclaves* (but not to be confused with *xenocrysts* which are inclusions of foreign mineral grains). Xenoliths which have elongate form and ill-defined boundaries are known as *schlieren* (although the term is also applied to similar features formed in different ways). They are found on all scales from minute millimetre-sized clots of dark crystals in a granite, up to masses of country rock tens of metres in size that have fallen into a magma body. Whereas it is easy to prove that small xenoliths are, in fact, surrounded by host igneous material, this may be less easy for apparent inclusions on the scale of tens of metres, where all contacts and boundaries are not exposed, and which might be parts of *roof* or wall rock separating different parts of an intrusion (see Chapter 5 for further details). Most xenoliths are recognized readily at the scale of a hand specimen within the otherwise uniform rock fabric. They comprise two main types: (i) those with relict meta-sedimentary textures and (ii) those with igneous textures.

(i) Xenoliths in the first category originated as fragments of country rock that were either engulfed by the upward-moving magma or fell into a magma chamber before crystallization became complete. Undisturbed sedimentary rock fabrics, such as stratification, may be preserved within these typically angular inclusions (Fig. 4.13a) but, more often, these xenoliths become severely metamorphosed by the heat of the magma, producing finely-banded hornfelsic textures. Xenoliths in lava flows erupted onto continental crust are particularly useful for studies of crust and upper mantle structure, as they may have sampled most of layers through which they have passed (e.g. upper mantle peridotite, high-grade metamorphic rocks of the lower crust, shales and limestones of the upper crust).

Fig. 4.13a Xenoliths in igneous rocks. A dark angular xenolith of hornfelsed slate in the Criffell granodiorite, Southern Uplands of Scotland.

Fig. 4.13b Partially melted xenolith of hornfelsed slate from the Shap granite, Cumbria. Note the leucocratic reaction rim surrounding this xenolith.

Fig. 4.13c Part of a coarsely-crystalline olivine nodule (light grey, representing olive green) in fine-grained basalt lava (dark grey) from Lanzarote, Canary Islands.

Country rock xenoliths engulfed by intrusive magmas often show evidence of reaction with the magma. At its most advanced stage, this results in partial assimilation of the xenoliths, leaving rounded dark-coloured inclusions, rich in recrystallized biotite and amphibole (Fig. 4.13b), with apparently igneous textures. Xenoliths in intrusive bodies (cf. Chapter 5) are often concentrated in number towards their source at the intrusion margins and so may be useful when mapping, especially as they may be aligned parallel to the intrusion walls.

(ii) Xenoliths of igneous origin: these range from the olivine and peridotite xenoliths, often called 'nodules', that are common in volcanic rocks (Fig. 4.13c) to fragments of pre-existing igneous bodies or of early-crystallized igneous material that became incorporated into the residual magma from the walls of a magma chamber (Fig. 4.13d). All igneous-textured xenoliths derived from a source in common with their host magma may be referred to as *cognate xenoliths*; the term *autolith* is appropriate for igneous xenoliths which represent earlier-crystallized igneous material from the same magma (e.g. Fig. 4.13d). Generally, however, these terms have genetic implications and should be used with caution until a full petrological and petrogenetic

Fig. 4.13d Autolith of microgranite in coarse-grained granite from Mt. Battock, Scotland.

51

analysis of the igneous body has been completed. Xenoliths of both igneous and sedimentary origin are particularly common in granites, granodiorites and diorites, and their study has major implications for the origin, evolution and mode of emplacement of such magmas (see, for example, Didier, 1973). In all cases, when studying xenoliths, evidence of a *reaction rim* surrounding the inclusion — in the form of changes in the mineralogy of the host within a few centimetres (typically) of the xenoliths — suggests that partial melting has occurred. In cases where xenoliths have been extensively melted and then recrystallized, they may develop igneous mineralogy and textures that are extremely similar to the host rocks; such inclusions are often known as *ghost xenoliths* (Fig. 4.13e).

Fig. 4.13e Ghost autolith of mesocratic syenite in granodiorite from the Quatre Tours granite, French Alps. Ghosting is due to partial dissolution of the autolith and chemical equilibration with the host magma.

Finally, as a general guide, *the colour, texture, mineralogy and margins of any xenoliths encountered should be described separately from their host rocks.* Further study of xenoliths, such as measuring their orientations and estimating their variability and the volume of the host rocks they occupy, will depend on the nature and objectives of the field work.

Table 4.9 Checklist for describing igneous rock hand specimens in the field (to be used in conjunction with Table 3.1 (p.25) for outcrop description)[1]

1 Examine weathered surface of rock outcrop, noting texture and colour: sometimes weathered surfaces provide a clue to the relative hardness of minerals and to their composition (e.g. the red-brown weathering residues of orthopyroxenes and olivines).

2 From the outcrop, collect and number representative sample(s) with fresh surfaces. When collecting for thin section or analytical work, trim off as much weathered material as possible in the field to save effort later and to lighten the load for carrying.

3 Record colour of the fresh surface and, where possible, estimate the colour index.

4 Examine the grains under a handlens:
   (a) If the rock is aphyric, note any other textural features and record felsic or mafic composition (see Table 4.2)
   (b) Record coarse, medium or fine grain size of the rock (see Table 4.3) and note textural relationship between minerals (Tables 4.4 and 4.5)
   (c) If the rock is porphyritic, record grain size and textural relationship of phenocrysts to groundmass.

5 Record the degree of homogeneity, the presence of layering, lamination, flow banding, vesicles and other special textural characteristics, such as the presence of inclusions (Section 4.6).

6 Estimate the number and proportions of the different minerals present and, for each, record where possible: colour, cleavage, lustre, habit, hardness. Use Table 4.6 for identification purposes.

7 Combine your observations to give the specimen a field name, using Figs 4.8–4.11 for medium- and coarse-grained rocks and Table 4.8 for porphyritic fine-grained rocks.

[1]See Fig. 4.12 (p.49) for an example of a field description following this scheme.

Mode of occurrence of igneous bodies

5.1 Volcanic rock units

Volcanic rocks are classified as *lavas* and *pyroclastic rocks*. Lava is the term for molten extrusive rock and its solidified product, and pyroclastic rocks are composed of materials fragmented by explosive volcanic activity. Pyroclastic rocks result from two types of deposit. *Air-fall* deposits result when pyroclastic material is erupted and falls back through the air to accumulate around the volcano. *Pyroclastic flows* result from transport of solid fragments of volcanic rock in a fluid (gas or liquid) matrix away from the volcano. Like lavas, such deposits are concentrated in valleys and depressions around the volcano. The most extensive pyroclastic deposits may form large-scale stratigraphic units that blanket the topography and may form plateau-like features around volcanoes.

Although lava flows are often regarded as the dominant product of volcanoes, *all* volcanoes erupt *both* lava *and* solid pyroclastic material. Whereas basaltic volcanoes such as Hawaii erupt dominantly (over 80%) lava, the dominantly andesitic products of many volcanoes in island arcs and active continental margins have less than 10% lava and over 90% pyroclastic rocks. Further, the erupted proportion of pyroclastic rocks is often underestimated from subsequent field studies because these materials are often rapidly dispersed by the wind, or are eroded after deposition more rapidly than the equivalent volume of solid lava. Hence, bear in mind that lavas might be over-represented in many island arc and continental margin volcanoes.

Volcanoes show a wide variety of forms, depending largely upon the composition of the erupted material and hence the style of eruption (cf. Chapter 6). During field investigation, keep in mind the likely scale of the volcanic form to which the studied rocks might belong. The most extensive volcanic areas comprise flat or gently-dipping *basalt lava plateaux*. These may have areas up to 10^5 km^2 and, since they occur within continental areas, they may be preserved within the geological record. Basaltic volcanoes are composed dominantly of lava and have a broad shield-like form, termed a *shield volcano*, which may have a diameter up to 100–200 km. These represent the largest discrete volcanic forms but, since they dominantly occur as oceanic islands, they are rarely preserved within the geological record.

Many volcanoes, particularly of andesite composition are *composite* in the sense that they comprise both lava and pyroclastic materials and have a steep irregular conical form. Such volcanoes are built by flow of lava down depressions around the volcanoes and the eruption of pyroclastic materials; they commonly have diameters of 10–40 km (Fig. 5.1). Although the lavas are generally restricted to the area of the volcano, the associated pyroclastic materials can be deposited at much greater distances from the volcano. For example, fine pyroclastic material may be transported by wind for a 1000 km or more. Such deposits (whether preserved within marine or continental sediments) may be stratigraphically correlated with an individual eruption by careful field and laboratory study. However, it may be difficult to correlate air-fall and flow deposits from the same volcano when these are dispersed in different directions depending, respectively, upon wind direction and topographic slope. Large-volume pyroclastic flows may form shield or plateau-like volcanoes within continental areas and are therefore more likely to be preserved than are basaltic shield volcanoes. Pyroclastic flows may travel up to *ca.* 100 km, and have volumes of up to 3000 km^3, so that shields or plateaux composed of such flow deposits may cover areas of 10^5–10^6 km^2.

The smallest volcanic forms result from a single short-lived eruption and comprise a variety of cones and extrusions. These include *pyroclastic cones* comprised of material of basic and intermediate composition, and steep-sided *flows* and *domes* of more viscous acid lava. Such volcanic forms are generally 1–2 km in diameter and may form upon, or near to, larger volcanic forms, when they may be termed *parasitic volcanoes*.

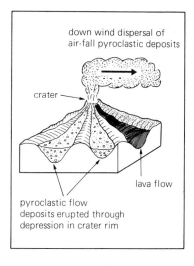

down wind dispersal of air-fall pyroclastic deposits

crater

lava flow

pyroclastic flow deposits erupted through depression in crater rim

Fig. 5.1 Forms of lava flows (shaded) and pyroclastic rocks erupted from a composite volcano.

From this description of volcanic forms and products, it is clear that a single volcanic area may be characterized by a variety of deposits. These may be contemporaneous and, for a terrestrial composite volcano, may be interpreted in terms of variation of *associations* of deposits with distance from the volcano (Fig. 5.2). The *central zone* (within *ca.* 2 km of the central vent) is characterized by lava conduits (later exposed as volcanic plugs, dykes and sills) associated with coarse, poorly-sorted pyroclastic materials which have been deposited near to the vent. The *proximal zone* (*ca.* 5–15 km from the central vent) has a higher proportion of lava flows, with a variety of pyroclastic flow deposits, and the *distal zone* (beyond *ca.* 5–15 km from the central vent, and extending beyond the volcano) is characterized by pyroclastic flow deposits associated with fine air-fall deposits dispersed by wind away from the volcano. These may be interbedded with sedimentary rocks such as lacustrine deposits. Pyroclastic cones, flows and domes may occur within any of the three zones. Therefore, even for young volcanoes, it may be difficult to correlate individual lava flows, pyroclastic fall and flow deposits with a single eruption.

The form of volcanic rock units is determined by study of the contacts. For lavas, the upper and lower contacts are generally irregular and the lower contact may be against the weathered surface of older rock. Where the lower contact is against an older lava flow, there may be an irregular mixture of rubbly material derived from the bottom of the overlying flow and the top of the lower flow. Lava flows are often thicker in areas of

low topography, and hence contacts between lava and older rock will be complex; in many cases, when traced laterally, the contact will be against a surface with considerable relief and will *not* represent a simple stratigraphic junction traceable over large distances. A further complication arises when lavas are erupted into valleys within a topography cut through older lava, in which case lava flows of different age may be exposed at a similar structural level (Fig. 5.3). For the more laterally extensive units of volcanic rock, age relationships are determined from simple principles of superposition, as applied to sedimentary rocks (see Tucker, 1982, Chapter 2).

In cases where fluid lavas (often basaltic) are repeatedly erupted over areas of low relief, extensive piles of *plateau lavas* result (Fig. 5.4). Sedimentary rocks such as lacustrine

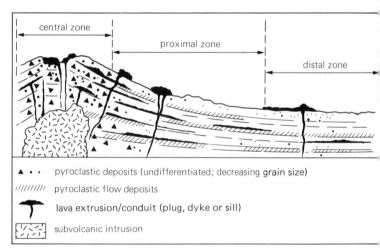

▲ • •   pyroclastic deposits (undifferentiated; decreasing grain size)

//////////   pyroclastic flow deposits

🌲   lava extrusion/conduit (plug, dyke or sill)

subvolcanic intrusion

Fig. 5.2 Schematic section showing the facies variations in volcanic rocks erupted from a large composite volcano. The characteristic features of the central, proximal and distal zones are described in the text.

deposits may be interbedded between the lava flows. Because of the large lateral extent of plateau lavas, these junctions may often be treated as regional stratigraphic contacts. Thus where the flows are nearly horizontal, the contacts will be close to topographic contours. However, in many cases, careful mapping shows that the thicknesses of lavas vary by more than is apparent in the field and care is needed to trace and measure individual plateau lava flows around the topography.

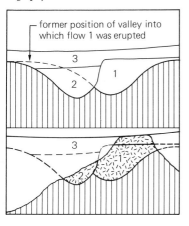

Fig. 5.3 Effect of differential erosion on erupted lava flows. Successive eruption of lava flows 1, 2 and 3 down a drainage channel has alternated with erosion along the margins of the lava flow. The broken lines indicate the former positions of the drainage channels. Following eruption of the earliest flow (1), later flows such as 2 may lie at a lower elevation, so producing an apparent stratigraphic order that differs from the true order of eruption. In the lower section on the left the apparent order (from the base) is 2–1–3.

Contacts between pyroclastic units may resemble those described for lavas above. The contacts of small-volume pyroclastic flows are again controlled by topography but large-volume pyroclastic flows, particularly ignimbrites (see Chapter 6), may be traced as stratigraphic units over large areas (Fig. 5.5). Their sharp upper and lower contacts may be traced and mapped using similar procedures to those applied to sedimentary rocks. Finally, remember that many pyroclastic rocks are redeposited by water (or wind) after formation. Hence, they show analogous contact relationships, depositional structures and facies changes to sedimentary rocks and may be mapped using the same procedures (see Tucker, 1982, Chapter 2).

5.2 Intrusive rock units

5.2.1 Minor intrusions

Intrusions vary widely in size and relationship to the country rock and are generally grouped according to size into

(i) *minor intrusions* which have mean minimum dimensions measured in tens of metres (or less) and were emplaced relatively near to the Earth's surface, and

(ii) *plutonic intrusions* which are commonly emplaced at greater depth and have sizes measured in terms of kilometers.

The most common forms of minor intrusions are shown schematically in Figs 5.6 and 5.7. *Dykes* (Fig. 5.6) are sheet-like intrusions which were approximately vertical at the time of emplacement and are hence *discordant* to host rocks such as shallow-dipping sedimentary rocks. As a consequence

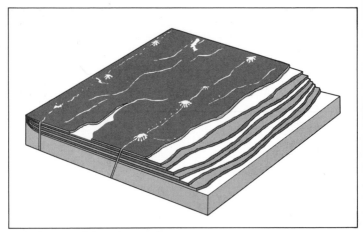

Fig. 5.4 Forms of plateau lavas (shaded) erupted from a fissure eruption.

Fig. 5.5 Ignimbrite plateau in north Chile. The ignimbrite, *ca.* 25 m in thickness, is exposed in a river gorge. In the background the composite volcano San Pedro rises to 6000 m, some 2000 m above the plateau, and a young lava flow (dark-coloured) has been erupted from a small parasitic pyroclastic cone at the base of the volcano.

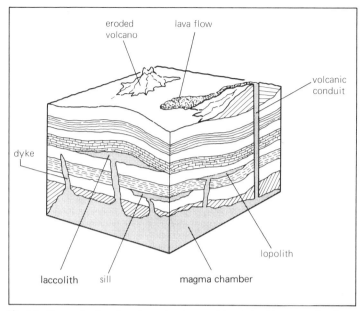

Fig. 5.6 Forms of minor intrusions, showing possible relationships to a sub-volcanic magma chamber. Sills, laccoliths and lopoliths are broadly concordant, while dykes are discordant tabular intrusions (see Figs. 7.1–7.4).

of their attitude, the outcrops of dykes are little affected by the topography of the country in which they occur and often appear as nearly straight lines on geological maps, maintaining a uniform direction over several and sometimes over many tens of kilometres. The width of dykes ranges from centimetre size (small dykes are more commonly termed *veins*, cf. Section 3.2) to sizes measured in hundreds of metres, but in general the average width is probably in the range 1–5 m. Since this is too small to portray accurately on large-scale geological maps, dykes are often shown with a uniform or 'conventional' width (like roads on large-scale geographic maps).

Sheet intrusions that were approximately horizontal at the time of emplacement are termed *sills*. These are often emplaced into horizontal or shallow dipping sedimentary rocks, in which case they may be broadly concordant with the stratification, and appear on a geological map as part of the sedimentary succession. Other forms of intrusion which are broadly concordant with the surrounding strata are blister-shaped masses with a sub-horizontal base and elevated upper surface, termed *laccoliths*, and saucer-shaped intrusions, termed *lopoliths*; both forms range in size from small bodies with dimensions measured in metres, to enormous masses hundreds of km in size, which are clearly distinct from smaller minor intrusions. Concordant intrusions may be mapped as stratigraphic units, but such intrusions may show rapid changes in thickness, and transgression from one unit to another (Fig. 5.6 and 7.4). Both dykes and sills may extend over areas measured in thousands of km² and sometimes show evidence of a relationship with surface volcanism.

Although most dykes are broadly linear features, some are related to central igneous complexes characterized by sheet-like intrusions that may be circular in outcrop plan. These are termed *ring intrusions* and are classified into two types according to the attitude of the contacts (Fig. 5.7). *Ring-dykes* have vertical and outward dipping contacts which are usually taken to indicate emplacement by *subsidence* of the central cylindrical block (Fig. 5.7a) in a tensional stress field. Ring-dykes may have thicknesses of up to 1–2 km but are more commonly in the order of several hundred metres. They may occur singly or in multiple units, and the overall ring complex may be 5–10 km in overall diameter. Cone-sheets occur as single small intrusions, with outcrops separated by curved sheets termed 'screens' of country rock. By contrast to *ring-dykes*, the form of cone-sheets must indicate emplacement by *uplift* of the central conical block and this indicates uplift of the roof of the chamber during intrusion. Ring-dykes are much more abundant than cone-sheets, but the latter are especially common in the Scottish Tertiary volcanic province. Because the mechanism of emplacement of ring-dykes involves subsidence of a central block (Fig. 5.7a), such ring intrusions are sometimes associated with relatively flat-lying

tabular minor intrusions emplaced over the block (Fig. 5.8). These include high-level granites, which are described in Section 7.3. Note that such sheet-like high-level intrusions are distinct in form from steep-sided plutonic intrusions.

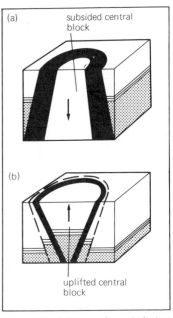

Fig. 5.7 Block diagram of (a) a single ring-dyke and (b) a set of cone-sheets, idealized to show the features of shape, outcrop and relationship to horizontally — bedded country rock.

In studying minor intrusions care should be taken to trace contacts and measure stratigraphic sections. Most contacts are complex: minor intrusions and plutons usually show steep contacts and, since the form and areal distribution of such contacts are used to determine the form of the intrusion in three dimensions, it is particularly important to determine the attitude and orientation of all planar contacts in order to determine the overall geometry. For many minor intrusions it should be possible to locate a contact area with sufficient three-dimensional exposure to determine dip and strike.

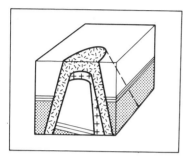

Fig. 5.8 Block diagram of two minor intrusions emplaced as a result of subsidence of a central block (cf. Figure 5.7a), idealized to show the features of shape, outcrop and relationship to horizontal country rock.

5.2.2 Plutonic intrusions

Coarse-grained plutonic rocks occur most frequently within large, elongate belts of intrusions, 50–150 km in width and 500–1500 km in length, that characterize eroded mountain belts. Such elongate plutonic bodies are termed *batholiths*. They are usually composed of a large number of cross-cutting smaller intrusions, including bodies 5–50 km in size with circular outcrop — literally *plutons* — and ring intrusions; the latter have steeply-dipping contacts. Circular intrusions with surface areas of less than 100 km² are sometimes termed *stocks*. The features of the intrusions in an eroded batholith are shown in Fig. 5.9. This

shows that the major contacts are outward dipping but that where the intrusion has an irregular roof, exposure might leave irregular downward projections of roof country rocks surrounded by igneous rock. These are termed *roof pendants* and can be identified from the outcrop pattern as shown on Fig. 5.9. Although the sub-surface form of plutonic intrusions may be determined from the attitude of contacts, additional clues come from the extent of the metamorphic aureole as explained below.

There are several lines of evidence that may be used to determine the orientation and attitude of a contact between a plutonic rock and country rock (see also Sections 3.2, 3.3). These include the characteristics of fine-grained contact zones, which are analogous to the chilled margin of a minor intrusion. Where a flow structure is visible in the igneous rock, whether as flow lines, preferred orientation of phenocrysts or mineral fabric, this might be taken as being parallel to a contact and hence may be used to infer its attitude. However, this method should be applied with caution since many mineral fabrics within intrusive rocks do not relate to contacts.

Plutonic rocks often have sets of joints at steep angles to the contact *and* parallel with the contact (Fig. 5.10) and, where these can be distinguished, then the jointing pattern may also provide a clue to contact orientation. Similar evidence may be obtained from the attitudes of inclusions and xenoliths (cf. Fig. 4.13). Where these show a preferential orientation that might reflect flow parallel to contacts, they may also be used to determine the attitude of contact (cf. Section 4.6). These features of intrusive rocks should all be considered when attempting to determine the position and attitude of contacts between plutonic rocks and their host rocks, which may consist of earlier-formed plutonic rocks.

In addition to features within the plutonic rocks, features of the country rock also provide clues about the attitude and orientation of the contact, for example, the extent of thermal metamorphism around a plutonic intrusion. Such metamorphism is broadly concentric with the contact of the intrusion and is hence termed a *metamorphic aureole*. The outer limit is drawn at the first detectable change in texture and/or mineralogy of the country rocks, and the degree of metamorphism increases towards the igneous intrusion (Fig. 5.9) over a distance usually of several hundred metres. The textures and metamorphic minerals that occur within metamorphic aureoles are summarized in Section 8.5. The contact metamorphic zone around an intrusion may vary from a few centimetres to many hundred metres, and reflects the size and temperature of the intrusion, the rate of cooling and the structure, composition and initial temperature of the country rock.

An example of a metamorphic aureole around a high-level granite intrusion is shown in Figure 5.11. Where an intrusion is emplaced into a rock which has stratification or cleavage planes, the metamorphic zone may be more extensive where the structure is at a high angle to the contact than where it is parallel (Figure 5.11a). This reflects the greater ease of transport of heat and hot gas *along*, rather than *across*, stratification or cleavage planes. Similarly, the contact metamorphic zone will be most extensive in limestones and shales, which are more

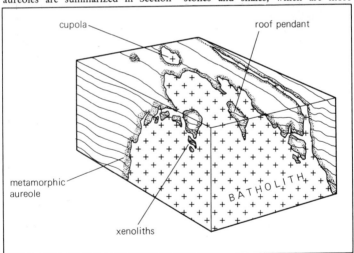

Fig. 5.9 Block diagram of part of a batholith (crosses) and its schistose country rock (lines). The stipple near the contact reflects the extent of contact metamorphism responsible for the metamorphic aureole.

susceptible to thermal metamorphism than sandstones. Where an intrusion is emplaced into country rocks of fairly uniform composition, the extent of the metamorphic zone may be used to infer the attitude of the contact, as shown in Fig. 5.11b. Hence, special attention should be given to the structure and composition of the country rock and the extent of the thermal metamorphic aureole.

5.2.3 General features of intrusive rocks

Contacts of both minor and plutonic intrusives are much more complex than those between volcanic rocks, and may be termed straight, jagged, blocky, sinuous, or diffuse (Fig. 5.12). Straight, jagged or blocky contacts may be interpreted as indicating the intrusion of magma into fissures in

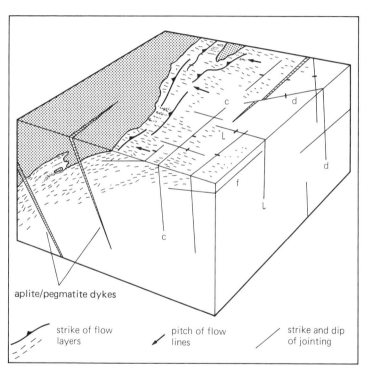

Fig. 5.10 Block diagram showing the relationship between flow structures and joint systems in an intrusive mass. c, cross joints; d, diagonal joints; f, flat-lying joints; l, longitudinal joints.

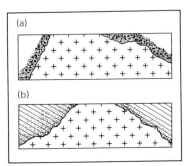

Fig. 5.11 (a) Section of an igneous contact between an intrusive rock (crosses) and its country rock. The latter is metamorphosed (stippling) over a wide area on the right, but this is simply a reflection of the shallow dip of the intrusive contact in comparison with that on the left.
(b) Section of an igneous contact between an intrusive rock (crosses) and stratified country rock (lined). The latter has been metamorphosed (stippling) in a wider zone on the left where the bedding is transverse to the contact.

brittle country rock that was cooler than the magma. For tabular bodies (Fig. 5.12d) the two contacts may have matching form, but often xenoliths are broken from the wall and surrounded by magma. More sinuous or lobate contacts indicate some plastic deformation of the country rocks during intrusion and hence imply that the country rock was warmer (possibly due to intrusion at a greater depth) than in the case of straight or jagged contacts. In some cases, contacts between intrusive rocks are diffuse (Fig. 5.12f) and this indicates that there was no chilling of one intrusion against another; both must have been at a similar relatively high temperature at the time of emplacement.

The determination of the *age relationships* of igneous rocks may be by direct or indirect procedures. The simplest method is the use of chilled contacts (see Figs 3.6, 3.10). Generally, younger igneous rocks emplaced into older rocks are chilled against them and the older rock may show evidence of contact metamorphism. However, in rare cases, evidence from chilling may be ambiguous; for example, emplacement of an igneous rock body into hot country rock can yield an unchilled contact. Similarly, emplacement of cooler acid magma into hot basic magma might form chilled contacts of basaltic against contemporaneous acid magma. Nevertheless, in spite of these complexities, the evidence of igneous contacts generally provides the clearest indication of relative age for igneous intrusions. Indirect methods for determination of age relationships include contact metamorphism, presence of xenoliths and the relationship of an intrusion to regional fault, joint and fold structures. Where the country rock and igneous intrusions show similar structures, this clearly indicates that the rocks predate the tectonic activity responsible for the regional structures. Conversely, many igneous intrusions show different and/or simpler structures than those of the surrounding country rocks, which might indicate a younger age for the intrusions.

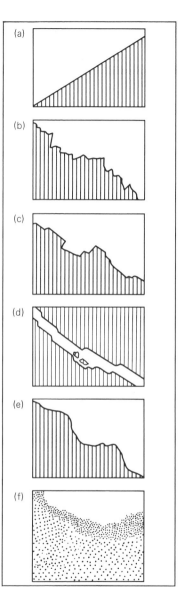

Fig. 5.12 Types of igneous contact, as seen in cross-section. The intrusive rock is blank and the country rock is lined or stippled (f). The contact is straight in (a), jagged in (b), blocky in (c), blocky and matched in (d), and sinuous in (e). (f) shows a diffuse contact (dark stipple) between an igneous intrusion (blank) and the country rock (light stipple). These contact relationships may be observed on scales ranging from centimetres to a few metres.

6
Volcanic rocks

Section 5.1 described the form and geological relationship of volcanic rock units; here we describe the *field characteristics* of volcanic rocks.

6.1 Lava flows

The most obvious features of lava flows are their form and surface morphology. The most extensive lava flows are basalts erupted from fissures, and such sheet-like flows with areas measured in terms of tens of km^2 form extensive lava plateaux (Figure 5.4). However, lava flows erupted over irregular topography, or from the summit or flanks of a shield or composite volcano, are tongue-like in shape, often filling depressions in the relief around the volcano (Fig. 5.1). On Hawaii, a typical basalt lava flow might be 10–20 km long, 200 m in width and 3 m in thickness, but there is great variation in such values because lavas vary in viscosity and rate of effusion. More siliceous flows have a greater viscosity and therefore tend to be relatively shorter and thicker; for example, andesitic lavas rarely flow more than 10–30 km from the volcano and many have thicknesses of up to 30 m. Few dacite or rhyolite flows travel more than 1–2 km from the vent; these may reach thicknesses of up to several hundred metres. Extrusions of dacite or rhyolite lava that are circular in plan and have a height : diameter (h/d) ratio of 0.5–0.3 are termed *domes* (Fig. 6.1). Domes commonly have heights of 50–

150 m and diameters of 150–500 m.

Lavas may be glassy, microcrystalline or fine-grained aphyric or porphyritic rocks (cf. Table 4.8). The chemical composition and mineralogy may be described using the principles explained in Chapter 4 but problems may be encountered as a result of the fine grain size. Thus, dark-coloured, fine-grained lavas may be difficult to classify in hand specimen. Completely aphyric lavas are rare and most lavas contain 10–50% phenocrysts, generally of a small range of minerals. An estimate of the lava composition can be made, and samples may then be given a field name based upon the abundant phenocrysts; for example, olivine-, plagioclase-phyric rhyolite. A summary of common phenocryst assemblages in lavas ranging from basalt to rhyolite in composition is shown in Table 6.1. In combination with features such as colour, field relationships and mode of occurrence, this Table is a useful guide to naming porphyritic rocks in the field. For example, olivine is most common in basalt and basaltic andesite lavas; abundant feldspar (plagioclase) with hornblende is characteristic of andesite; and quartz and feldspar (alkali feldspar) are diagnostic of rhyolite. However, rhyolites are more commonly aphyric than lavas of basic and intermediate composition.

The approach taken above can be applied only to volcanic rocks that are

Table 6.1 Composition of major phenocryst phases in porphyritic lavas

	Basalt	Basaltic andesite	Andesite	Dacite	Rhyolite
Plagioclase	**	***	***	***	**
Olivine	**	**	*	—	—
Pyroxene	**	**	**	*	—
Hornblende	*	*	**	**	*
Biotite	—	—	*	**	**
Alkali feldspar	—	—	*	**	***
Quartz	—	—	—	**	***
Fe-Ti oxide	**	**	*	—	—

*** often present   ** frequently present   * rarely present   — absent (or rare)

sufficiently medium- or coarse-grained for the individual mineral grains, or at least the phenocrysts, to be seen and identified using either the naked eye or a handlens. More rapidly cooled rocks are fine-grained, microcrystalline or glassy, in which case they may be difficult to identify, but can have additional distinctive features.

Basic rocks composed of natural glass are termed *tachylite* and these form on the skin of lava flows or along the marginal zones of basic sills or dykes. Volcanic glass of intermediate and acid composite is termed *obsidian* or *pitchstone*. Obsidian is a black or dark-coloured glass, with a bright glassy lustre and well-developed conchoidal fracture. Pitchstone has a dull, *resinous* or pitchy lustre, rather than the glassy lustre of obsidian. (This reflects the high content of water in pitchstone in comparison with obsidian.) Fragments of both types may occur as pyroclastic materials, or form minor intrusions or lava flows. Glassy volcanic rocks of intermediate and particularly acid composition commonly show flow banding (cf. Fig. 3.11) as a result of the drawing out of small inhomogeneities during viscous flow.

Such natural glasses are unstable and start to hydrate and to crystallize or *devitrify* soon after formation. This process takes place progressively so that, in general, there are hardly any unhydrated natural glasses older than *ca.* 20 Ma, and, although glasses of Palaeozoic age are known, such glasses rarely exceed 100 Ma in age. However, many ancient volcanic rocks must have originated as natural glasses (cf. flow-banded rhyolite in Fig. 3.11). Crystallization occurs by spontaneous growth of tiny crystals or *crystallites*, commonly of quartz and feldspar. Usually, the individual crystals are not visible in hand specimens but devitrification crystals, in radiating clusters forming small circular structures resembling cod roe or oolites (termed *spherulites*), may be scattered throughout the devitrified rock. Complete devitrification yields an aphyric microcrystalline rock, which may be light-coloured lava. (Such rocks may be termed *felsite*, but the term is also applied to any light-coloured fine-grained rock composed chiefly of quartz and feldspar.)

6.1.1 Surface morphology

The surfaces of lava flows may be described as belonging to *four* major types (Fig. 6.2 and 6.5). *Pahoehoe flows* (Fig. 6.2a) have smooth, billowing

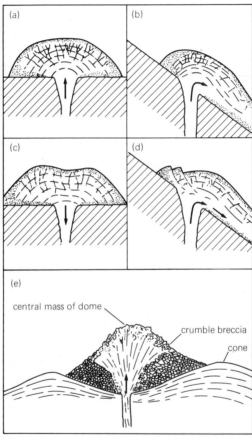

central mass of dome

crumble breccia

cone

Fig. 6.1 Schematic cross-sections of domes, showing flow lines (fine lines), joints (heavy lines), and brecciated margins (stippled). The arrows indicate the direction of lava movement. (a) Dome built on a nearly horizontal surface. (b) Dome built on a sloping surface. (c) The top of the dome has sagged due to drainage of magma back into the vent and to slight spreading of the dome. (d) Summit of the dome broken by faults due to down-slope movement of the main mass of the dome. (e) Cross-section of a dome built within a composite cone, showing steep fan structure developed within a confining wall of crumble breccia.

ropy or entrail-like crusts of quenched glass. Ropy crusts are sometimes considered characteristic of pahoehoe flows but are much less common than smooth crusts and generally *only* occur in mafic (basic) lavas. *Aa flows* (Fig. 6.2b) have rough clinkery and spine-like (spinose) surfaces reflecting tearing apart of fragments of the solidified crust on the top of lava that is more viscous than that forming pahoehoe flows.

(a)

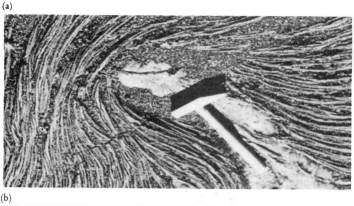

(b)

Fig. 6.2 The principal forms of lava surface, (a) pahoehoe, Hawaii, U.S.A. Length of hammer is 35 cm. (b) aa, Mount Etna, Italy.

Both pahoehoe and aa lava surfaces occur on mafic lavas and may occur upon the same lava flow, changing from pahoehoe near to the vent to aa further downslope. Pahoehoe lava surfaces may, therefore, form on hotter, less viscous and hence faster flowing lavas than aa surfaces. Whereas aa surfaces occur on both basalt and basaltic andesite lavas, most andesite, dacite and rhyolite lavas are characterized by *blocky lava* surfaces (Fig. 6.2c). These surfaces have detached polyhedral blocks with planar or slightly curved faces and conspicuous dihedral angles.

Lava flows commonly show evidence of vesiculation. Vesicles may be perfect spheres, but this is rare; they are more usually ellipsoidal or almond-shaped, evidence for the direction of lava flow. After formation, gases and hot solutions circulating through the rock may deposit minerals within vesicles. Such deposition may continue by circulation of groundwater long after the lava is cold. The mineral infillings are termed *amygdales* and these rocks may then be termed *amygdaloidal* (Figs. 6.3). The commonest amygdale minerals include calcite, silica and zeolites; cf. Table 4.6.

Fig. 6.2c Blocky lava, San Pedro volcano, North Chile.

Such amygdaloidal fillings may be partial or complete and the filling may be zoned, as shown in Fig. 6.3, reflecting progressive deposition of hydrothermal minerals toward the centre. In older terranes flattening of amygdales might occur as a result of tectonic deformation accompanying folding.

This may be distinguished from flow flattening, as the former should be parallel to regional tectonic trends on a medium (i.e. outcrop) and larger scale.

Structures formed during movement and solidification of basic lavas can be used to determine the stratigraphic top, and the overall shape and direction of movement within a flow. Certain associations of structure may be so characteristic of a given flow that they may be used for tracing units and mapping contacts. An idealized section through a pahoehoe basalt flow is shown in Fig. 6.4. Non-viscous basic lavas may show alignment of phenocryst minerals such as plagioclase (cf. Section 3.3). Such lavas generally contain vesicles which may occur within continuous lines or sheets, giving the rock a banded appearance or platy fracture, but such structures are *not* necessarily parallel to the overall flow direction. Larger vesicles are sometimes more abundant in the upper parts of flows and may be stretched out parallel to the last movements of the lava. Within the lower part of the flow, cavities formed by upward movement of gas have a tubular form and are termed *pipe vesicles* (or amygdales). The occurence of such vesicles may therefore be used to locate the base of the flow and possible directions of final movement of the lava. Withdrawal of lava from the centre of a solidified lava flow may leave a hollow space within the flow which is termed a *lava tube*. These are common in pahoehoe flows and the occurrence of filling material on the sides or base of the tube may give evidence for the position of the top of the flow. The interiors of most aa flows are more massive than those described above (cf. Fig. 6.4), but some have a platy structure defined by sheets of small vesicles. Vesicles may be aligned in complex flow patterns but these are less well developed than in the case of pahoehoe lavas described above.

During the advance of an aa or blocky lava flow, the solid surface materials are continuously over-run by the fluid body of the lava. Thus a sequence of successively erupted sequences of lavas will comprise alternations of solid lava separated by material characteristic of the surfaces. (The lava flows may also be separated by other materials such as pyroclastic

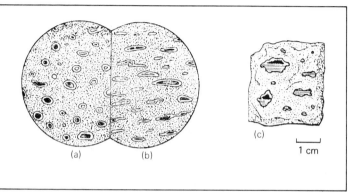

Fig. 6.3 Amygaloidal structure in lava. (a) and (b) Silica (white) was deposited on the walls of the original vesicles, and subsequently chlorite (lined) filled the remaining open spaces. The vesicles in which the amygdales of (b) were deposited were flattened and drawn out by the flowing of the lava. (c) Amygdales and larger cavities with lining and stratified fillings of silica.

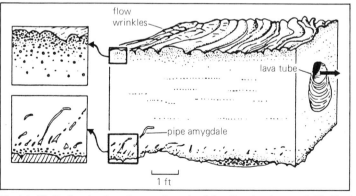

Fig. 6.4 Idealized section through a non-viscous basalt lava flow showing a partly filled lava tube, wrinkled and vesicular pahoehoe surface and pipe amygdales near to the base.

71

or sedimentary rocks.) Where there is a short interval between the flows, the contact will be of rubbly material with no sharp contact between one flow and the next. Such extrusions have been termed *compound lava flows*. Where there is a longer interval, the top of the lower flow may show effects of subaerial weathering. Where this weathering has taken place under tropical conditions, the lava surface may show surface oxidation, and in the case of mafic lavas the surfaces may be converted to a deposit of red-brown hydrated Fe- and Al-oxides, known as *laterite*.

Pillow lavas (Fig. 6.5) are basalts that erupted below water and, since the upper part of the ocean crust is basaltic, such lavas may be the most abundant volcanic rocks of the Earth's surface. However, it should be emphasized that pillows may form in any depth of water below a few metres, and even in mud and in melt water below a glacier. They are formed when lava is chilled rapidly by water: a chilled rind forms on the surface which is sufficiently flexible to move with the flow which remains molten inside. But as lava pressure increases, the rind breaks, causing formation of a propagating bulbous or tubular lobe. Such pillows vary from a few centimetres to several metres in their greatest dimension. The pillows have thin skins composed of basaltic glass, *tachylite* (reflecting rapid chilling of lava entering sea water), which in some cases has been altered by oxidation and absorption of water into a yellowish-brown waxy-looking substance, termed *palagonite*, and then to low-grade metamorphic minerals (e.g. zeolite and chlorite). Palagonite may form subaerially or below water. The pillows are sometimes connected by slender necks and have tops that are usually convex upward, whereas the undersurfaces may be flattish, concave upward or project downwards into interstices between the underlying pillows (Fig. 6.6a). These characteristics may be

Fig. 6.5 Pillow lavas, Troodos ophiolite complex, Cyprus. Length of hammer is 35 cm.

used to infer the way-up of the volcanic pile.

The interstices between the pillows are often filled with fragments of broken glassy lava skins, cherts, limestone or shale, which when compacted, are lithified and later converted to metamorphic minerals. Many pillows have a radiating columnar structure and may show concentric zonation of vesicular zones, produced as volcanic gases were episodically exsolved within the core. However, since the pressure of sea water at a depth of 3000 m is sufficient to prevent exsolution of dissolved gases, the vesicularity of pillows varies from *ca.* 10–40% when they formed at shallow depths (less than 500 m) to below *ca.* 5% at 1000 m (Fig. 6.7), and this may be used as a crude estimate of the depth of water into which the pillow lavas were erupted.

Pillows may be spheroidal, ellipsoidal or so flattened (even when undeformed tectonically) that they may be difficult to distinguish from other similar structures. For example, many ancient basaltic lavas contain pillow-like structures, such as lobes of (subaerial) pahoehoe lava, that might be confused with true pillow lavas (Fig. 6.6). In view of the palaeogeographic significance of pillow lavas (particularly in the identification of ophiolite complexes, see Chapter 10) and their value as 'way-up' indicators, you should examine the outcrop carefully for the features described above.

A further product of eruption of basaltic magma into water results from large-scale fragmentation of glassy lava (e.g. pillow rinds and pillow lava fragments). Such fragmentation produces a deposit composed of glassy basaltic particles of sand size mixed with separated pillows. Such deposits are termed *hyaloclastite* (Fig. 6.8), or *palagonite-tuff* when the basaltic glass has been palagonized. Hyaloclastite commonly

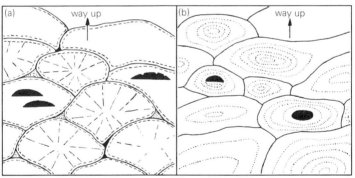

Fig. 6.6 Diagrammatic cross-section showing the distinction between pillow lava (a) and lobes of pahoehoe lava (b). In (a) the short dashed lines represent the inner margin of the glassy skin on the pillows, the long dashed lines represent radial cooling joints, and the stippled areas represent sedimentary material filling interstices between the pillows. In (b) the dotted lines reflect concentric variations resulting from cooling of the lava lobes. In (a) and (b) the black ornament represents spaces which would be open in young lavas, but would be filled with secondary material in older lavas.

forms massive poorly-bedded deposits associated with subaerial lava, or submarine pillow lava (Figure 6.8). In some cases hyaloclastite overlain by lava may be interpreted in terms of advance of (subaerial) lava from a shoreline into water to build a lava 'delta' that eventually emerges above water level. Sometimes the hyaloclastite shows distinct bedding (cf. Fig. 6.8) and this might then reflect reworking of the fine-grained glassy material by waves and currents, so that the deposits may contain sedimentary structures reminiscent of sand deposited in shallow water.

6.1.2   Internal jointing

Many basaltic lavas, sills and dykes are characterized by *columnar jointing* (Fig. 6.9). These reflect contraction during cooling of solidified lava. These joints are generally orientated perpendicular to the upper and lower cooling surfaces of the flow, and split the lava into polygonal columns. The columns typically have 5 or 6 sides but some have 3 or 4, and a few may have 7 sides. The columns are mostly straight and parallel-sided, but some curve and vary in width when traced over a distance exceeding a few metres.

Columnar jointed lava flows often show a three-fold division (Figure 6.10a), with an upper and lower unit of straight-sided well-developed columns, termed *colonnades*, separated by a middle unit composed of blocky, irregular columns arranged in a complex way, termed the *entablature*. However, in some examples the entablature may be relatively thin, in which case the upper and lower colonnades may resemble two separate lava flows, and other examples may comprise a thick entablature overlying a single

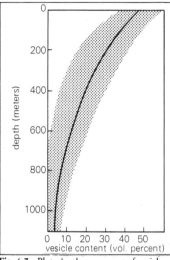

Fig. 6.7 Plot of volume percent of vesicles against depth in metres for submarine basalts from south of Iceland. The solid line is a best fit line drawn through the field of data (stipple); below 1000 m, the vesicularity is less than *ca.* 5%.

colonnade (Figure 6.10b).

Intermediate and acid lavas sometimes show columnar jointing, but more commonly develop *platy* or *slabby joints*. The lower massive parts of andesite and dacite flows may be characterized by abundant close-spaced platy joints, ranging in dimensions from 1 to 50cm, parallel to the underlying floor or curving down against the direction of movement, i.e. flat lying in the original lava (Fig. 6.11). Traced upwards, such joint planes tend to become more widely spaced and slabbly, and become more nearly vertical or even overturned near to the surface. Steep-sided platy joints on the surface of dacite or rhyolite extrusions may be termed *ramp-structure*.

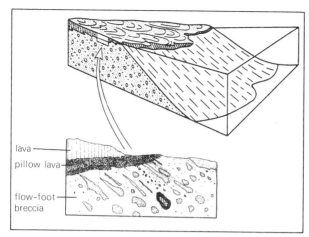

Fig. 6.8 Cross-section of a pillow lava delta formed by basalt eruption associated with brecciation of lavas as they come into contact with water. Below the pillow lava, the delta consists of bedded hyaloclastite ('flow-foot breccia') containing isolated elongate pillows within the plane of bedding.

Fig. 6.9 Columnar jointing in the Giants Causeway lava, County Antrim, Ireland.

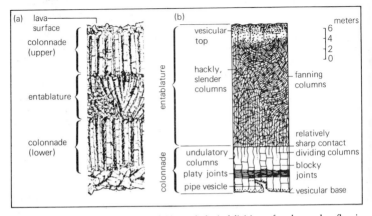

Fig. 6.10 (a) Three-fold, and (b) two-fold morphological divisions of a columnar lava flow in terms of colonnades and entablature.

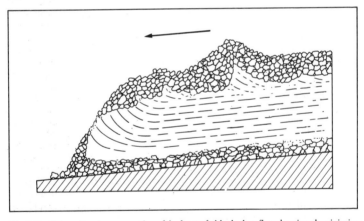

Fig. 6.11 Diagrammatic cross-section of the front of a blocky lava flow showing platy jointing (broken lines). The arrow indicates the direction of movement of the flow. The flow thickness may be 5–20 m.

6.2 Pyroclastic rocks

Pyroclastic rocks are composed of materials fragmented by explosive volcanic activity. Three types of fragment may be found in any pyroclastic deposit.

(a) *Fragments of new lava* which are termed *juvenile* fragments and which may range from solid unvesiculated material to fragments of highly vesiculated lava.

(b) *Individual crystals*, yielded by liberation of phenocrysts present in the juvenile lava as a result of fragmentation.

(c) *Lithic fragments* which may include any older rock within the deposit, but often comprising older lavas.

6.2.1 Problems in the study of pyroclastic rocks

Although pyroclastic rocks are locally abundant within the geological record, there are many problems in their study and interpretation. For instance:

(a) Field identification of rock type may be difficult. For example, fine-grained lavas and tuffs may be difficult to distinguish, and brecciated lava surfaces may be confused with agglomerates. In these cases, mapping the form of the body may be diagnostic but ambiguities frequently may remain.

(b) Although many pyroclastic rocks described above are subaerially deposited, these are readily distributed by water and such undisturbed deposits are more abundant in ancient sequences than are subaerial deposits. Such reworked pyroclastic rocks may be characterized by structures indicating deposition within water (e.g. Tucker, 1982, Chapter 5).

(c) Pyroclastic flows may be emplaced within a submarine environment and in such cases may occur between layers of marine sediments, and may grade up into such sediments (which may occasionally contain marine fossils). Air-fall pyroclastic deposits erupted onto water settle out as discrete ash layers, resembling those described earlier (Section 5.1) but often showing greater grain-size vari-

ation between top and bottom due to slower settling of fine ash through water, and may show reverse grading of pumice due to temporary flotation, before settling, of larger fragments.

Such ash layers occur within modern oceanic sediments over wide areas of the ocean floor and can be recognized within older marine sedimentary rock successions. However, emplacement of such pyroclastic materials within the zone of abundant wave- and current action, leads to partial or total reworking. Such reworking forms shallow water sedimentary structures.

6.2.2 Classification of pyroclastic deposits

Pyroclastic deposits may be classified in two quite different ways. A *genetic classification* relates the deposits observed with inferred eruption processes and so would be readily applied to relatively young volcanic deposits. Observations would include mapping variations in deposit thickness, performing accurate grain-size analyses and mapping internal structures. Where the characteristics of ancient deposits have been obscured by metamorphic and tectonic processes, it may be impossible to use such an approach, in which case a *lithological classification* is more appropriate. This relies on description of the major characteristics of the deposit; these may then be used to infer the mechanism of eruption and to apply the genetic classification.

Pyroclastic fragments are classified by the type of material and the grain-size limits shown in Table 6.2. Details of the classification procedure are discussed after considering the constituent materials of pyroclastic rocks. The components of a typical air-fall ash are shown in Fig. 6.12.

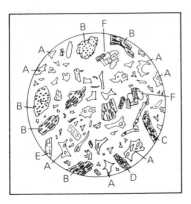

Fig. 6.12 The components of a typical air-fall ash of dacite composition. The diameter of the field is about 2 mm. The ash has a fine matrix which has been omitted and consists of angular glass shards (A), some showing cuspate forms; pumice fragments (B) showing vesicles that are nearly round in cross section in one direction and much stretched in the direction right angles to the first; and crystals of hornblende (C), biotite (D) and feldspar (E) with a little quartz, and lithic fragments (F).

Juvenile vesiculated fragments (Fig. 6.12A, B) range from highly vesiculated pumice (density may be less than 1000 kg m^{-3}), through less well vesiculated juvenile fragments, sometimes termed 'expanded andesite or dacite' (density *ca.* 1000–2000 kg m^{-3}) to poorly vesiculated or non-vesiculated rock fragments (density over 2000 kg^{-3}). *Pumice* is generally taken to apply to fragments of intermediate or acid chemical composition of density *ca.* 1000 kg m^{-3} or less. *Scoria* is the term used for dense vesicular fragments of basic or intermediate composition. The term *glass shard* is applied to ash-sized fragments. (Table 6.2) resulting from the fragmentation

of pumice or scoria walls, so tending to have Y-shaped or cuspate shapes (Fig. 6.12A).

Crystals may be derived by fragmentation of partially-crystallized porphyritic magmas and often may behave as a distinct population during transport and deposition within pyroclastic deposits (Fig. 6.12 C–E).

Lithic fragments (Fig. 6.12 F) are usually the densest components of a pyroclastic deposit and include non-vesiculated juvenile fragments which are genetically related to the volcanic host, termed *cognate lithics*, and accidental country rock fragments incorporated during eruption or transport and which are, respectively, termed *accessory* and *accidental lithics*.

6.2.3 Genetic classification of pyroclastic rocks

There are four main types of pyroclastic deposits:

(i) *Pyroclastic fall deposits* resulting from explosive eruption of pyroclastic material from a volcanic vent into the atmosphere. Such eruptions range in scale from those responsible for formation of small volcanic cones (*scoria cones*, cf. Fig. 6.13) to air-fall deposits that blanket the flanks of larger composite volcanoes and the surrounding ground. During a large eruption the volcanic plume containing fragments expands and the pyroclastic materials fall back downwind at varying distances from the vent. Pyroclastic fall deposits have many structures that resemble those of other sedimented strata. Individual beds show a uniform thickness over areas that lie on elongated contours downwind around the vent, draping all the steepest topography with *mantle bedding* (Fig. 6.14). The deposits are generally *well-sorted*, show *normal* grading of pumice and lithic fragments, and may show *internal stratification* in grain size or composition, due to variation in the force and composition of the eruption.

(ii) *Pyroclastic flow deposits* result from lateral movement, close to the ground, of pyroclastic fragments, travelling as a hot *high-concentration* gas/solid mixture. The deposits are topographically-controlled, filling valleys and other depressions and covering the lower elevations around the volcano (Fig. 5.1). They are *poorly-sorted* (but may show coarse-tail grading; a negatively-skewed grain-size distribution resulting from selective removal of fine-grained material, leaving a 'tail' of coarser material in a histogram) and generally *lack internal stratification*. Pyroclastic flows may show normal grading of lithic fragments but *reverse-grading* of pumice fragments, reflecting their lower density in comparison with the transporting gas/solid pyroclastic flow. The superposition of thin, inter-

Table 6.2 Grain size limits for pyroclastic fragments (after Fisher, 1961)

Grain size[1]		Pyroclastic fragments
(φ)	(mm)	
−8.0	256	coarse } blocks and bombs[2]
		fine }
−6.0	64	
		lapilli
−1.0	2	
+4.0	1/16	coarse } ash
		fine }

[1]The grain size is expressed in two scales. The 'phi' (φ) scale has values which decrease arithmetically as grain size (expressed in mm) increases geometrically. The conversion from the mm scale to the phi scale is arithmetically very simple and is expressed by the equation $N = 2^{-\phi}$, where $\phi = 1, 2, 3$, etc. and N is the grain size in mm.

For example, consider a grain size of 2 mm:

$$N = 2^{-\phi}$$
$$2 = 2^{-\phi}$$
and $$2 = 1/2^{\phi}$$

Therefore $2^{\phi} = \frac{1}{2} = 2^{-1}$
So, for a grain size of 2 mm, $\phi = -1$.

Therefore, zero on the phi scale corresponds to a grain size of 1 mm, whereas $\phi = +1$, $+2$ and $+3$ correspond, respectively, to $\frac{1}{2}$, $\frac{1}{4}$ and $\frac{1}{8}$ mm and $\phi = -1, -2$ and -3 correspond, respectively, to 2, 4 and 8 mm, etc.

[2]A block that takes on a distinctive form during flight or upon landing is termed a volcanic bomb. Such bombs are often given descriptive names such as ribbon bombs, spindle bombs, breadcrust bombs and cowdung bombs.

nally poorly-sorted pyroclastic flow units may easily be confused with stratification of air-fall deposits (which are, nevertheless, usually much better sorted) or with other sedimentary rocks.

The texture of pyroclastic flow deposits may become modified during compaction. This is most marked in *pumiceous pyroclastic flows* which are termed *ignimbrites*. The major process is *welding*, which is the term for post-depositional sintering of hot vesicular fragments (including pumice) and glass shards during compaction. Welding requires the presence of glassy material, and is therefore characteristic of pumiceous pyroclastic flows (although many ignimbrites are unwelded). The term *welded ignimbrite* is analogous to *welded tuff* (or welded *ash-flow tuff*). Ignimbrites in which welding is most fully developed show zones characterized by dense welding, partial and incipient welding, and no welding (Fig. 6.15). In the zones of dense welding, the glass shards and larger flattened pumice fragments (*fiamme*) define a planar foliation termed *eutaxitic texture* (Fig. 6.16). Incipiently welded tuffs, in which there is no textural evidence of flattening of glassy constituents, are termed *sillar*. During welding, lithic fragments resist deformation, so that the hot, plastic glassy material may be deformed around them.

(iii) *Pyroclastic surge deposits* involve lateral movement of pyroclastic fragments as a hot *low-concentration* gas/solid mixture. The deposits tend to blanket topography, but are more concentrated in valleys and depressions than air-fall deposits. Characteristically, they show unidirectional sedimentary flow forms such as cross-stratification, dune structures, planar lamination, anti-dunes and pinch-and-swell structures (Fig. 6.17). Therefore although characterized by internal stratification, the individual laminae are generally well-sorted. There is complete gradation between high concentration pyroclastic flows and low-concentration pyroclastic surges, with consequent gradation between the geological characteristics described. As for air-fall deposits, pyroclastic flow and surge deposits may be erupted from small cones, and from larger composite volcanoes (cf. Figs. 5.1, 5.2).

(iv) *Lahars* Pyroclastic flows form a continuum between high temperature flows (over 100°C) in which the pyroclastic material is transported by a gaseous phase (these are the flows described above) and low temperature flows. Below 100°C, the transporting medium is a mixture of liquid water and gas. Fragmented volcanic materials deposited from a cool water-rich flow may be referred to as *mudflows*. The deposits frequently consist of large volcanic fragments in a finer matrix which is commonly coarser than mud-grade and is ash-grade material (Table 6.2). The Indonesian term *lahar* is preferable for such water-deposited poorly-sorted deposits of volcanic materials. Besides showing a transition to hotter pyroclastic flows, lahars are also variable in terms of the proportion of juvenile and non-juvenile volcanic material. Eruptions of volcanic material through lakes, below ice or during heavy tropical rain may generate lahars composed largely or entirely of juvenile volcanic material. Conversely, non-juvenile material may be erupted and there may, therefore, be a transition from pyroclastic lahars to water-rich *debris flows* formed on the slopes of volcanoes. These are especially common where deposits form in shallow-water conditions subject to volcanic

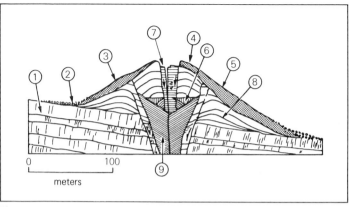

Fig. 6.13 Cross-section of a scoria cone built on a surface of lava flows and pyroclastic rocks (1) exposed on the flank of a composite volcano. The cone is surrounded by a ring of coarse blocks, commonly up to *ca.* 1 m in diameter (2), and the surface is covered with fine blocks (3), lapilli and ash (4) and unsorted ash and blocks (5). The vent may contain relicts of ponded lava (6) subsequently buried or destroyed by collapse or large eruptions from the vent (7). The interior of the cone (8) comprises similar but earlier-erupted deposits and the vent may be clogged with inward-dipping pyroclastic materials (9).

Fig. 6.14 Mantle bedding of air-fall ash, Irazu, Costa Rica. The thick layer of ash, showing faint bedding, is *ca.* 30 m in thickness and overlies older well-bedded pyroclastic deposits.

81

instability and seismic activity. These include the cold lahars described above, which commonly contain angular blocks of partially lithified fragments of a uniform pyroclastic rock type. Such deposits do *not* result directly from volcanic activity but reflect disruption and flow of partially lithified pyroclastic material. Such material may not be strictly volcanic in origin and hence may be termed sedimentary-slump deposits (cf. Tucker, 1982, Chapter 5).

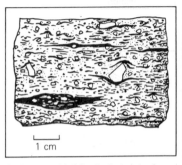

Fig. 6.16 Welded rhyolite ignimbrite showing black glass lenses formed by flattening of pumice lapilli, compacted around larger lithic fragments, and abundant crystals (cf. Tucker, 1982, Fig. 3.14).

At outcrop, lahars with abundant juvenile material may resemble hot pyroclastic flows as described above, but may be distinguished from them by a greater degree of heterogeneity and a greater content of large lithics and boulders, and the absence of any evidence of high temperature features such as welding, thermal alteration, carbonized organic material and irregular pipe-like structures from which fine ash has been removed as a result of upward streaming. Such structures are termed 'fossil fumarole pipes'. Finally, lahars, like other pyroclastic flows, are concentrated in valleys and topographically low ground around a volcano. Many lahars are 10–20 km in length and some are known to have travelled over 300 km from their source.

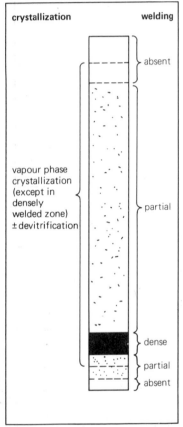

Fig. 6.15 Variation in degree of welding within a single ignimbrite unit. The section is *ca*. 70–100 m in thickness.

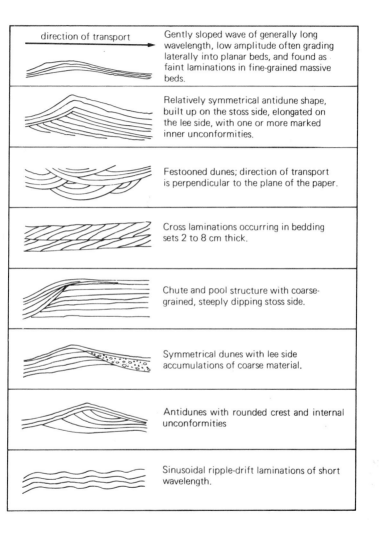

direction of transport →	Gently sloped wave of generally long wavelength, low amplitude often grading laterally into planar beds, and found as faint laminations in fine-grained massive beds.
	Relatively symmetrical antidune shape, built up on the stoss side, elongated on the lee side, with one or more marked inner unconformities.
	Festooned dunes; direction of transport is perpendicular to the plane of the paper.
	Cross laminations occurring in bedding sets 2 to 8 cm thick.
	Chute and pool structure with coarse-grained, steeply dipping stoss side.
	Symmetrical dunes with lee side accumulations of coarse material.
	Antidunes with rounded crest and internal unconformities
	Sinusoidal ripple-drift laminations of short wavelength.

Fig. 6.17a Morphologies of structures commonly found in pyroclastic surge deposits.

Fig. 6.17b Pyroclastic surge deposit (cf. Tucker, 1982, Fig. 3.15).

6.2.4 Lithological classification of pyroclastic rocks

The basis of the lithological classification adopted is:

(i) The grain-size limits of the pyroclasts and the grain-size distribution of the deposit.

(ii) Composition of the pyroclastic fragments.

(iii) The degree and type of welding.

(i) *Grain size* The terms used for pyroclastic fragments within different grain-size limits are shown in Table 6.2. The grain-size characteristics may be determined from a cumulative curve of the grain-size distribution, from which the median diameter Mdϕ (= ϕ 50) and $\sigma\phi$ = (ϕ84 − ϕ16)/2 which represents the graphical standard deviation and a measure of sorting is derived. Grain-size analysis of unlithified and non-welded pyroclastic deposits is used to differentiate between fall and flow deposits on the basis of their degree of sorting (Fig. 6.18). This demonstrates that pyroclastic flow deposits usually have $\sigma\phi$ values generally greater than 2.0 and are poorly sorted in comparison with pyroclastic fall deposits, which have values of less than 2.0. Flow deposits are usually dominated by ash-sized particles (less than 2 mm in grain size; Fig. 6.18) and are often termed *ash-flows*, yielding *ash-flow tuffs* on lithification. (An ash-flow tuff is therefore a type of ignimbrite, cf. p.80. The pyroclastic fall- and flow fields tend to overlap on plots of Mdϕ against $\sigma\phi$ (cf. Fig. 6.18), indicating that individual flow laminae may reveal the relatively poor sorting of the flow deposit. Clearly, an important aspect of study of an unlithified pyroclastic deposit is its grain-size analysis; a crude qualitative estimate of these characteristics from a thin section or photograph is a useful indication of the mechanism of deposition.

(ii) *Constituent fragments* The components present in pyroclastic deposits may be readily distinguished in young, unlithified or poorly lithified pyroclastic deposits. They may be difficult to identify in finer grained lithified deposits. An important effect of lithification is to infill vesicles in pumiceous material and pore spaces in scoriaceous material. However, the major components such as infilled pumice fragments and glass shards may be distinguished in older, lithified deposits.

(iii) *Welding* The style of welding (if present) should be examined for variations in degree. The features of welding zonation shown in Fig. 6.15 may be recognized in older ignimbrites on the large scale of a rock face, at the medium scale of a hand-specimen and on the small scale of thin section. Welding is most common in pyroclastic flow deposits but can also occur in fall deposits.

6.2.5 Pyroclastic rocks erupted through water

Eruption of magma into, or through, water results in another suite of characteristic volcanic products which include the *hyaloclastites* (described in Section 6.1.1). The more violent pyroclastic eruptions in shallow water form small low-profile cones ($h/d = 0.1$–0.2) composed of fine-grained ash-grade pyroclastic material termed *ash-* or *tuff-rings*. Similar pyroclastic material is formed by eruption from larger composite volcanoes having crater lakes, or during rainfall. In comparison with subaerial pyroclastic fall deposits, such material is generally finer-grained (Md generally below 2 mm) and less well-sorted ($\sigma\phi$, 1.5 >). The individual layers are generally thin, commonly 1 mm to 1 cm, in comparison with much coarser subaerial pyroclastic layers (layers *ca.* 1–5 cm). Characteristic features of pyroclastic material erupted through water are:

(i) *Accretionary lapilli* (Fig. 6.19), which are small spheres varying in size from 1 mm to several centimetres, but commonly 2–10 mm in diameter. They may consist of a relatively thin shell of fine-grained ash around a core of coarse ash, or be composed wholly of concentric layers of fine-grained ash (as in oolites). Such lapilli are believed to form by accretion around solid nuclei in a medium of condensed moisture (cloud and/or rain) within an eruption column.

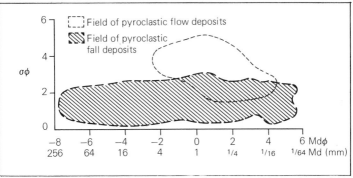

Fig. 6.18 Plot of standard deviation, $\sigma\phi$ against median diameter, Mdϕ showing the fields of pyroclastic flow and fall deposits.

(ii) *Impact structures*, commonly termed *bomb-sags* (Fig. 6.20), which result from fall of bombs (or more commonly, *blocks*) into finely bedded, water-saturated sediments. Such impact structures are characteristic of pyroclastic deposits erupted in wet conditions.

Many pyroclastic successions show parts of a characteristic stratigraphic sequence of deposits (Fig. 6.21). This starts with a coarse air-fall deposit, and is followed by a pyroclastic surge, a pyroclastic flow (which may be welded to form ignimbrite) and lastly a fine pyroclastic fall deposit. Each ignimbrite flow unit (one is shown in Fig. 6.21) has a fine-grained basal layer (2a) ranging in thickness from a few centimetres to a metre, overlain by the main layer (2b), up to several tens of metres in thickness, that shows the characteristic normal grading of lithic fragments and reverse grading of pumice fragments described earlier. This layer comprises over 90% of the ignimbrite flow unit and is overlain by a fine air-fall ash deposit (3). Although this sequence has been described from many volcanic areas, you should bear in mind that it does not apply to all volcanic areas, which may show evidence for different sequences or part sequences of eruption.

Both pyroclastic fall and flow deposits may show evidence for chemical and mineralogical (compositional)

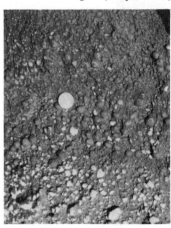

Fig 6.19 Accretionary lapilli (dark spheres) in air-fall ash deposit (containing light-coloured fragments). The lapilli are *ca.* 3–10mm in diameter.

Fig. 6.20 Impact structure ('bomb sag') in thinly-bedded basaltic ash deposit. The block is *ca.* 20 cm in diameter.

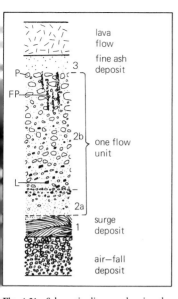

Labels on figure:
- lava flow
- fine ash deposit
- one flow unit
- surge deposit
- air—fall deposit

(with numbers 3, 2b, 2a, 1 and letters P, FP, L marking the column)

Fig. 6.21 Schematic diagram showing the textural elements of a complete pyroclastic eruption episode. An inversely graded air-fall bed is overlain by a pyroclastic surge deposit (1) showing some of the structures illustrated in Fig 6.17a. The basal layer of the pyroclastic flow (2a) may show inverse grading, whereas the main part of the flow (2b) may have normal grading of lithic fragments and reverse grading of pumice fragments. Therefore, lithic fragments (L) are concentrated near the base, and pumice fragments (P) are concentrated towards the top. Fumarolic pipes (FP) may be present throughout the flow. Deposits of fine ash (3) from the cloud would occur above the flow unit. A lava flow might cap the sequence and help to preserve it in the geological record.

zonation, and magma mixing. *Compositional zonation* results from eruption of a compositionally-stratified magma within the magma chamber such that, for example, the first erupted juvenile material may be more silicic and less mafic than the last erupted. In this case, the resultant pyroclastic deposit will then have a lower portion that is more silicic than the upper portion. The change may be gradual throughout the sheet or abrupt within it and the composition represented usually ranges between varieties of andesite and dacite, although more extreme variations between basalt and rhyolite are known. Many pyroclastic flows also show an increase in phenocrysts towards the top of the sheet, implying a downward increase in phenocryst content in the magma chamber. Further evidence of stratification in the magma chamber and/or magma mixing is provided by pumice fragments that contain intimately mixed material of contrasting composition: these are termed *mixed pumices*. However, some caution is necessary, since many pumice fragments which appear to contain mixed material owe their appearance to differing degrees of vesiculation or devitrification and may be uniform in chemical composition.

6.3 Summary of the field characteristics of volcanic rocks

1 Volcanic rocks comprise lava flows and pyroclastic rocks. Lavas have homogeneous or flow-banded crystalline textures in contrast to pyroclastic rocks, which have fragmental textures and may show internal stratification and/or depositional structures. Further, lava flows and pyroclastic deposits may have distinctive three-dimensional forms which may be distinguished by field mapping (see below).

2 Lava flows range in form from extensive plateau-like flows of non-viscous basalt, through tongue-like

flows of andesite and dacite, to steep-sided extrusions of more viscous dacite and rhyolite composition. Lavas have distinctive surface textures and internal jointing patterns. Pillow lavas result from eruption of basalt lava below water.

3 Pyroclastic rocks result from fragmentation of magma by volcanic activity and comprise fragments of juvenile lava, crystals and lithic fragments. They may be divided into fall and flow deposits, which range from high-temperature flows to low-temperature lahars and debris flows.

4 Fall deposits are well-sorted, may show internal stratification and blanket the topography, whereas flow deposits are more poorly-sorted, generally lack internal stratification, and fill valleys and depressions around the volcano, or form large volcanic shields or plateaux. Pyroclastic surge deposits are characterized by distinctive depositional structures and some pumiceous pyroclastic flows (termed ignimbrites) may show internal variations, including development of a welding fabric.

The forms of minor intrusions were introduced in Chapter 5. These intrusions may be conveniently divided into linear-tabular and circular-globular groups and the forms of these groups are summarized below. The linear-tabular group includes dykes and sills, both commonly associated with tensional environments such as constructive plate margins (Section 1.2). Circular and globular minor intrusions include cone-sheets, ring-dykes, laccoliths and lopoliths and also volcanic plugs and diatremes (cf. Section 7.2). The emplacement of ring-dykes and cone-sheets may be related to ring faulting associated with the formation of craters and calderas (a caldera is a crater exceeding 1 km in diameter), and plugs, whereas diatremes apparently represent the infilled vents of extinct volcanoes. These minor intrusions are widespread within many volcanic areas and are not especially associated with any single tectonic environment.

7.1 Dykes and sills

The commonest form of minor intrusions are termed *dykes* and *sills* (Figs. 5.6 and 7.1). A *dyke* (US, dike) is a tabular intrusion that cuts across horizontal or gently-dipping planar structures, such as stratification, in the surrounding rock. It therefore has the same form as a vein (described in Section 3.2). A *sill* is a tabular intrusion that parallels such horizontal or gently-dipping planar structures. Dykes are therefore *discordant* to structures in the country rock whereas sills are generally *concordant* between beds of layered rock but may show small side transgressions and/or larger dyke-like transgressions (Fig. 7.1).

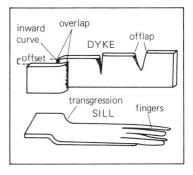

Fig. 7.1 Idealized three-dimensional forms of dykes and sills. See text for discussion.

These definitions raise complications; in deformed terrains it may be difficult or impossible to determine whether a given tabular intrusion was emplaced as a dyke or as a sill (however, extensive sills are commonly emplaced only within well-stratified sedimentary sequences). A variety of minor intrusions is known, and these are described in terms of their relationship to the country rocks. Many extinct and eroded volcanic areas are characterized by the occurrence of dykes and sills and, in such cases, these minor intrusions may have been intimately connected with surface volcanism and may have channelled the magmas that were erupted. The connection between minor intrusive activity and surface volcanism is not clear cut; minor intrusions are described below.

Minor intrusions have mineral compositions corresponding to most other igneous rock types. They range from basalt through andesite and dacite to rhyolite in composition and may be named after the volcanic equivalents (cf. Table 6.1). Thus a fine- or medium-grained minor intrusion of mafic mineral composition may be termed a basalt, although the terms *dolerite* and *diabase* are also used. These terms imply a particular intergrown textural relationship between plagioclase and pyroxene (observable only in thin section and termed ophitic texture) and are synonymous in US usage; however, in the UK the term diabase is sometimes used for a dolerite in which the feldspars and mafic minerals are highly altered. Intermediate and acid dykes are termed fine-grained diorite, granodiorite and granite, and, if they are aphanitic, may be termed andesite, dacite and rhyolite as described for lava flows. Glassy aphanitic minor intrusions may be termed pitchstone or obsidian as defined in Section 6.1. Although most minor intrusions may be matched in composition with common rock types described earlier, a rather uncommon group, the lamprophyres (Table 4.7) which are distinct in mineral composition and texture from normal plutonic and volcanic rocks, form minor intrusions (see Chapter 4).

Dykes (Fig. 7.2) usually vary from a centimetre to many metres, or even hundreds of metres, in width but large elongated discordant mafic intrusions with width measured in terms of kilometres may also be termed dykes. Dykes that radiate away from individual igneous centres are termed *radial dykes*, and, where such dykes (and those that cannot be related to such centres) are concentrated parallel to a regional tectonic trend, they are termed *dyke swarms*. When traced along strike, an individual dyke is rarely continuous. It may show discontinuity at outcrop, termed *offlap*, and may locally be offset in such a way that parts of the dyke show *overlap* (Fig. 7.1).

Fig. 7.2 A dyke cutting sedimentary rocks on the south coast of Ardnamurchan, Scotland. The dyke has a sinuous outcrop pattern and has prominent wide-spaced joints normal to the margins. Width of dyke = *ca.* 1.5 m.

Fig. 7.3 A rhyolite sill (thickness = *ca.* 6–8 m) showing crude columnar jointing (below castle), intruded into an older basic intrusion (base of cliff) cutting gently-dipping sedimentary rocks (foreground).

Sills (Fig. 7.3), like dykes, range from a few centimetres to hundreds of metres in thickness, although the commonest sills are between 1 and 20 m in

thickness (cf. Chapter 5). Frequently they underlie large areas, up to several hundred km² of country (Fig. 7.4). They may show attenuated extremities which separate into elongate *'fingers'* (Fig. 7.1). Unfortunately, sills may be easily confused with basaltic lava flows (especially where both occur in the same sequence). However, they may be distinguished from flows: first, by the absence of any of the surface characteristics of lavas (as described in Chapter 6), and, second, by the occurrence of fine-grained chilled margins (against the country rocks) at the lower *and upper* contacts.

In addition to dykes and sills, ring intrusions (Section 5.2.1) are locally abundant. These include *ring-dykes*, which have steeply-dipping contacts and range up to 5–10 km in diameter. Their field characteristics may therefore encompass those of minor and plutonic intrusions as described here and in Chapter 8. When emplaced close to the surface, ring-dykes may be associated with sheet-like high-level intrusions and the characteristics of these are described in Section 7.3. *Cone-sheets*, in contrast, are normally only 1–2 m in thickness and hence,

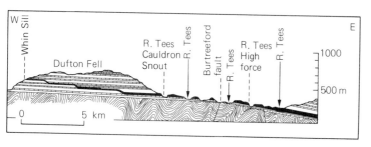

Fig. 7.4 Cross-section through the northern Pennines (UK) to show the location of the Whin Sill within gently-dipping Carboniferous strata (brick ornament = limestone; plain = shale and sandstone) unconformably overlying folded Lower Palaeozoic (Silurian) rocks.

apart from their form, resemble dykes and sills within the field (Fig. 7.5). Ring-dykes and cone-sheets may be medium- or coarse-grained in texture and are more commonly of mafic rather than felsic mineral composition.

For minor intrusions, it is important to study the igneous margins and the contact with the country rock (cf. Chapter 5). Contacts of minor intrusions are usually chilled and hence are fine-grained or glassy in character. They may provide a clue to the state and composition of the magma emplaced to form the minor intrusions; for example, the volume and composition of the phenocryst phases indicates the proportion and type of crystals present within the magma at the time of intrusion. The grain-size variation across the intrusion may provide information concerning the cooling of the intrusion; a large variation in grain-size implies emplacement of hotter magma into cooler country rocks than does a lesser degree of variation. A margin that is *not* chilled implies that the intrusion was emplaced into warm or hot country rocks. The country rocks around minor intrusions frequently show the effect of *thermal metamorphism*. For most minor intrusions, this is limited to within a metre, or at most a few metres, of the contact. The metamorphic recrystallization and the minerals in the contact rocks should be noted; for example, shales may be baked and hardened, limestone may be recrystallized to contact metamorphic marble, and rocks rich in organic material may be carbonized. Details of such metamorphism are given in Chapter 8, Table 8.3.

The relative ages of different groups of minor intrusions may often be determined from contact relationships. Thus, if more than one kind of minor intrusion occurs, for example as discerned by composition and/or geographic trends, then the outcrops should be studied for intersections, as shown in Fig. 7.6. In such cases several episodes of injection by the same kind

Fig. 7.5 Cone-sheets forming small transversely-jointed minor sheet-like intrusions, cutting Jurassic sedimentary rocks on the south coast of Ardnamurchan, Scotland. The cone-sheets dip at angles of *ca*. 20–30° towards an igneous centre (Centre 2) to the northwest (i.e. to the top left).

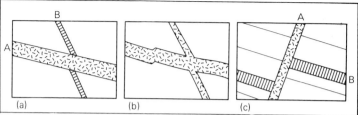

Fig. 7.6 (a) Intersecting dykes, A being younger than B. Note that the apparent displacement of B results only from dilation by emplacement of A, and *not* from lateral movement within the younger dyke, (b). Branching dykes, of the same age. (c) section of dyke, A, which has entered a fault. The fault interesects a series of strata containing a sill, B.

of minor intrusion may be demonstrated (cf. Fig. 3.6). However, care should be taken not to confuse intersecting dykes with branching dykes (Fig. 7.6). Where contacts are not exposed, the relative ages of minor intrusions may sometimes be determined from identification of xenoliths of one type within another. Also, minor intrusions can be related to regional tectonic activity, such as folding and faulting, and the relative ages

should also be determined (Figs. 7.6, 7.7). The emplacement of sills into well-bedded strata may also lead to complex field relationships which may easily be misinterpreted. For example, it is sometimes easy to confuse transgressions by sills with faulting (Fig. 7.7). The absence of offsets in the beds above and below the sill (Fig. 7.7b) confirms that this represents a simple transgression, whereas Fig. 7.7c shows a sill (and the enclosing rocks) offset

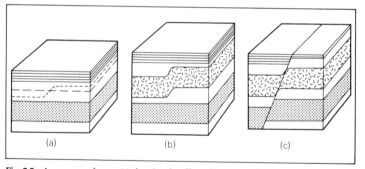

Fig. 7.7 A sequence of strata (a) showing the effect of intrusion of a locally transgressive sill (b) or a completely concordant sill which is later displaced by faulting (c). The position of the transgressive sill (b) is shown by the short dashed lines in (a). In each case, the final disposition of the sill is similar (compare (b) and (c)) but the displacement of strata above and below the sill and the possible identification of deformation (e.g. fault brecciation) along the fault plane enables (c) to be distinguished from (b).

by faulting. More complex situations result, for example, from intrusion of a sill (A) followed by intrusion of a second sill (B) (Fig. 7.8).

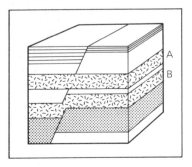

Fig. 7.8 Sections of sills that have been emplaced into faulted strata. A is a sill that has been emplaced across the fault plane, whereas B is a sill emplaced at a particular stratigraphic horizon and which exhibits transgression at the fault plane. The simple case A may be interpreted by its discordant relation to the faulting and the transgressive case is distinguished by the lack of internal deformation within the sill (cf. Fig. 7.7b).

Although most minor intrusions result from injection of a single pulse of magma, successive intrusion of magma into the same intrusion may lead to formation of more complex minor intrusions. If new magma of similar composition is emplaced into the still-molten interior between the chilled margins of an earlier intrusion, this is termed a *multiple intrusion*. The formation of multiple dykes may be favoured by emplacement at sufficient depth for cooling to be sufficiently slow for dykes to be partially molten at the time of emplacement of the next intrusion. A particular association of multiple dykes termed a *sheeted complex*, forms part of ophiolite com-

plexes, and these are described in detail in Chapter 10. In rarer cases, a multiple intrusion might be composed of two contrasted magma types. These are termed *composite dykes* or *sills*. The commonest case is when an early intrusion of basic magma has been followed by an intrusion of acid magma (Fig. 7.9); the contact between the two types varies between very sharp and completely diffuse. In composite intrusions, the basic and acid components might contain xenoliths or xenocrysts of each other. *Homogeneous* minor intrusions might contain inclusions of contrasted composition, for example basalt or andesite inclusions within dacite or rhyolite, which indicate magma mixing as outlined for pyroclastic rocks.

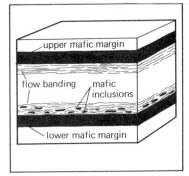

Fig. 7.9 Diagrammatic section of a composite intrusion (black = mafic, white = felsic).

In addition to compositional variation in the main rock-type(s) within minor intrusions, variations in the type and distribution of phenocryst phases can be important. For example, phenocrysts may be concentrated in the centre of a minor intrusion which has fine-grained aphyric margins (Fig.

7.10), reflecting *flowage differentiation* in which phenocrysts concentrated in the centre of the intrusion reflect more rapid magma flow compared with that at the margins. However, such cases might also reflect composite intrusion of aphyric and porphyritic magma, so it is important to examine carefully the texture of possible contact zones in such heterogeneous intrusions in order to distinguish between these possibilities. Evidence for a sharp (or chilled) contact between the inner (porphyritic) intrusion would favour interpretation by multiple intrusion, although the magmas may have been related by fractional crystallization below the depth of emplacement.

Besides the variations described above, sills, in particular, show other types of mineralogical variation as a result of their horizontal tabular form. These include variation as a result of

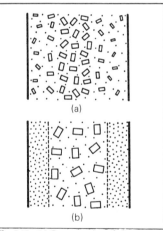

(a)

(b)

Fig. 7.10 (a) Concentration of phenocrysts in the centre of a dyke as a result of flowage differentiation. (b) A composite dyke with an aphyric margin sharply separated from a porphyritic centre.

crystal settling (cf. Section 3.3) which occurs to a noticeable extent within some thick sills, generally exceeding 50 m in thickness. Such sills may have a layer or layers rich in dense minerals such as olivine, pyroxene and Fe-Ti oxides near to the base (cf. Chapter 10). These layers overlie the fine-grained lower marginal layer and may be complemented by layers relatively poor in these minerals towards the top of the sill, below the upper chilled margin. However, it may be extremely difficult to distinguish between such mineralogical differentiation of a single magma, and that resulting from multiple or composite intrusion. For example, fine-grained olivine and/or pyroxene-rich dolerite may intrude aphyric dolerite. Distinction of such origins may require detailed chemical and mineralogical study of the sill.

As a result of fractional crystallization and volatile concentration during crystallization, many sills are characterized by segregations of coarser-grained and/or more felsic material in the upper part. These include sharp-sided veins, irregular segregations and irregular diffuse streaks termed *schlieren*. They vary from fine-grained aplites through medium-grained (including granophyre) to coarse-grained texture and may have large crystals (up to 10–20 cm long), particularly of pyroxene, amphibole and Fe-Ti oxide. They are then termed *pegmatitic schlieren*. The composition of segregations in sills generally reflects the composition of the crystallizing magma. Felsic segregations in quartz-bearing basalt/dolerite sills are fine-grained diorite, granodiorite and granophyre whereas those in quartz-free, olivine basalt/dolerite sills are often quartz-free gabbro or diorite, containing zeolites and/or foid minerals.

7.2 Volcanic plugs and diatremes

Erosion of composite volcanoes may reveal subcircular exposures of fine-grained rocks that may be interpreted as *volcanic plugs*. Such plugs are also termed *volcanic rocks*, or *volcanic pipes*, and are generally a few hundred metres in diameter, but may extend up to a kilometre. The three-dimensional form of volcanic plugs is generally cylindrical. Most plugs are composed of lava, but a small proportion consist of pyroclastic material; these are described separately below. The lava within plugs may be of various grain sizes and compositions; it often shows alteration and deposition of secondary minerals as a result of hydrothermal activity concentrated around the cooling volcanic vent. Most plug rocks show features such as brecciation (i.e. an unsorted mixture of angular fragments) as a result of the passage of gases and hydrothermal solutions through them (Fig. 7.11)). Such brecciated rocks are termed *volcanic*
agglomerate or *breccia*. Many small pyroclastic volcanoes are underlain by plugs or pipes composed of pyroclastic material. These are termed diatremes, and include *breccia-pipes*, *tuff pipes* and *kimberlites*. Diatremes are narrow, steep-sided cylindrical or funnel-shaped pipes which have intruded through the crust, and consist of solid fragments of crustal and mantle rocks with or without solid fragments of juvenile rock, apparently emplaced at a low temperature. Diatremes may be composed mainly of comminuted crustal rock or juvenile material. Some diatremes have erupted at the Earth's surface as *ash-* or *tuff cones*; others can be shown not to have reached the Earth's surface but to have been revealed subsequently by erosion, and are then termed 'blind pipes'. The crustal fragments in diatremes range from fine ash to blocks several metres in size. Juvenile fragments are generally glassy ash and pumice but may include crystalline material.

Fig. 7.11 Volcanic vent breccia Greenscoe vent, English Lake District. Width of field of view = *ca*. 50 cm.

In natural exposures of diatremes, the walls of the vent are usually visible over only 100–200 m; most diatremes seem to converge downwards like steep-walled funnels. The rocks forming the diatreme may be bedded as a result of infall of products of successive eruptions and frequently they can be matched with rocks in the surrounding walls. Where diatremes were emplaced within a well-defined stratigraphic sequence, matching of such fragments may indicate their upward or downward movement. Commonly the fragments appear to have *subsided* with the diatreme with layers of tuff as shown in Fig. 7.12. Such subsidence may be up to several hundred metres (or, rarely, 1500 m) in vertical extent. The tuff and breccias comprising such diatremes are termed *tuffisite* and *intrusive breccia*, respectively, to distinguish them from extrusive rocks.

A distinctive rock that occurs in diatremes and occasional dykes is *kimberlite*, which is of economic interest since many kimberlites contain diamonds. Kimberlite is a serpentinized and carbonated mica-peridotite of porphyritic texture, containing nodules (i.e. xenoliths) of ultrabasic rock-types characterized by high-pressure minerals such as pyrope (garnet) and jadeitic diopside (pyroxene). They are dark-green to dark-blue grey, and are often extremely brecciated with large crystal or rock fragments within a fine-grained matrix. Kimberlite pipes include an extraordinarily varied assortment of xenolithic inclusions; for example, angular blocks of the country rock, rounded blocks of high-grade metamorphic rocks, such as might form the lower continental crust, and rounded blocks of ultramafic rocks composed of varying combinations of olivine, pyroxene, garnet

(pyrope), spinel, Fe-Ti oxide and mica (phlogopite). These rocks may be named according to the scheme shown in Fig. 4.11 and are believed to represent samples of the upper mantle.

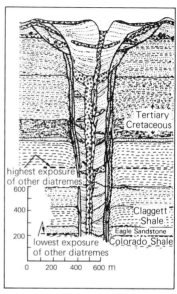

Fig. 7.12 Schematic composite diagram of a typical diatreme based on examples that have been eroded to various levels in north-central Montana.

Many kimberlites are altered by hydration and/or carbonation. Indeed, to judge from the presence of carbonate alteration around some kimberlite vents (such as at Fen in South Norway, Fig. 9.3), carbon dioxide seems to have been a common gaseous component of these solid-gas diatremes. The internal structure of a kimberlite pipe is usually so chaotic as to defy systematic description, so a comprehensive field study would normally consist of mapping the pipe boundaries and collecting a representative suite of rock samples.

7.3 High-level subvolcanic intrusions

Although some minor intrusions, and many plutonic intrusions, show no clear geological relationship with surface volcanic activity, many eroded volcanic areas reveal small intrusions of fine-grained diorite, granodiorite and granite composition that were probably subvolcanic in character and which may be termed *high-level intrusions*. Such intrusions may be approximately circular or elliptical in outcrop with a diameter of 1–4 km, or might form arcuate outcrops that may have been emplaced by subsidence of an overlying block as part of a ring intrusion (cf. Fig. 5.8).

High-level subvolcanic intrusions are characterized by association with contemporaneous volcanic rocks, sharp chilled contacts (cf. Fig. 5.12) and a rarity of features such as pegmatites and aplites that are generally associated with intrusions at deeper levels. They may also have small cavities up to several millimetres in size lined by inward projecting euhedral crystals of the host intrusion, or of hydrothermally deposited crystals such as quartz, feldspar and zeolites. These are interpreted as vapour cavities, analogous to amygdales, and are termed *druses* or *drusy cavities* (Fig. 7.13). Rocks containing such cavities are termed *miarolitic rocks*. The presence of drusy cavities indicates exsolution of a fluid phase under low confining pressure and hence provides evidence of crystallization shallower than a depth of *ca.* 1–2 km.

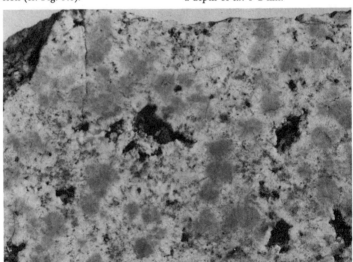

Fig. 7.13 Drusy cavities in a hand specimen of granite from the Island of Skye (the Loch Ainort granite). Width of field of view = 10 cm.

High-level intrusions may contain the range of xenoliths and inclusions described in Section 4.6 and these are sometimes more conspicuous near the upper margins than in deeper-level intrusions. In particular, high-level intrusions often contain sharp-sided angular xenoliths which are absent in deep-level plutonic intrusions. It is often difficult to establish the type and degree of connection of high-level intrusions with the surrounding volcanic rocks, and therefore careful study should be made of geological features such as xenoliths, inclusions and contact characteristics that might contribute to the establishment of such relationships.

7.4 Summary of the field characteristics of minor intrusions

1 Minor intrusions may be divided into linear-tabular and circular-globular groups.

2 The commonest types of minor intrusions are linear-tabular in form and include dykes and sills. A dyke is a tabular intrusion that cuts across horizontal or gently-dipping planar structures, such as stratification, whereas a sill is a tabular intrusion that is broadly parallel to such planar structures.

3 Minor intrusions of circular-globular form include ring-dykes and cone-sheets. Ring-dykes are circular in plan with steeply-dipping contacts consistent with emplacement by subsidence of a central block, whereas cone-sheets have inward-dipping contacts, indicating emplacement by uplift of a central block.

4 Minor intrusions may be emplaced by successive injection of magma of similar composition (termed *multiple* intrusions) or by magma of contrasted composition (*composite* intrusions), and some intrusions show compositional evidence for differentiation *in situ* by processes such as crystal settling and melt segregation.

5 Circular intrusions of lava or pyroclastic rock, respectively, may represent volcanic plugs or diatremes, which in turn represent the infilled vents of extinct volcanoes. Circular or tabular intrusions of fine-grained igneous rock associated with volcanic rocks may be intrusions emplaced at a high level (high-level intrusives) below an area of active volcanism.

Plutonic rocks I: the calc-alkaline association

8.1 Introduction to Chapters 8–11

Plutonic rocks in the calc-alkaline association and in most other categories form intrusions emplaced at depths of several kilometers (cf. Chapter 5) and are composed of medium- to coarse-grained rocks (cf. Chapter 4). The form of plutonic intrusions is illustrated schematically in Fig. 8.1, which makes the point that the relative amounts of intrusive rock and country rock available for study depend strongly on the level of erosion. The fabric of plutons is more complex than that of extrusive associations; often it is very complex and irregular in form, but sometimes regular linear or planar fabrics are developed (cf. Chapter 3), and sometimes these involve alignment of phenocrysts, megacrysts or inclusions (xenoliths), see Figs 8.2 and 8.3. Whereas such structures commonly reflect magmatic processes (e.g. convectional alignment within a magma chamber), quite similar structures can be developed when plutonic rocks are subject to post-consolidation deformation. At low pressure–temperature conditions, such tectonic processes might be responsible for a wide or close-spaced fracture cleavage in plutonic rock (Fig. 8.4). Along cleavage planes, the crystals may be broken,

deformed or, at higher temperatures, recrystallized to yield a planar fabric resembling that resulting from magmatic flow. At higher pressure–temperature conditions, metamorphic fabrics are more diffuse and the fabric may be difficult to interpret in terms of purely magmatic or purely metamorphic conditions. Under the highest pressure–temperature metamorphic conditions, plutonic rocks can recrystallize as metamorphic rocks, such as gneisses; these are described in Chapter 12.

Sufficient general information relating to the study of outcrops, field specimens and the form of plutonic intrusions appears in Chapters 3–5; the object of this and the following three chapters is to pay special attention to examples of and particular problems of field study for plutonic intrusions, which are grouped into four subcategories by mode of occurrence and association of rock types: the calc-alkaline association (this chapter); the alkaline association (Chapter 9); the mafic-ultramafic association (Chapter 10), and the anorthositic and charnockitic associations (Chapter 11). We begin with the calc-alkaline association.

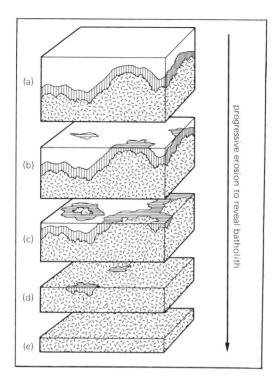

Fig. 8.1 Successive stages in the exposure of batholith rocks (dash pattern) revealed by erosion of country rocks (white) and the metamorphic aureole (straight lines).

8.2 General features and occurrence

Most of the rocks in the calc-alkaline association are *granitoids*, i.e. they are granites, granodiorites and tonalites all containing >20% quartz, with feldspar, mica and amphibole in various combinations (Figs 4.7 and 4.8); they occur together with subordinate amounts of diorite and gabbro. In recent years, there have been renewed

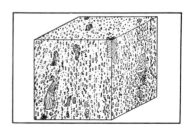

Fig. 8.2 Linear parallelism of phenocrysts, mineral streaks and xenoliths in an intrusive body.

101

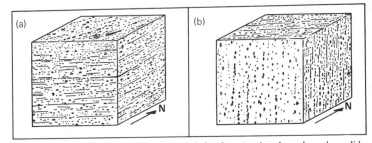

Fig. 8.3 Platy parallelism of phenocrysts (tabular shapes), mineral streaks and xenoliths (showing internal structures) arranged in parallel layers, (a) horizontal parallel flow layers with NE-SW oriented flow lines and (b) vertical flow layers and lines.

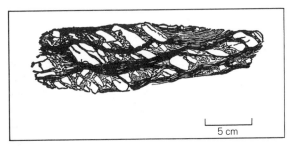

Fig. 8.4 Sketch of sliced feldspars in a sheared igneous texture.

attempts to classify calc-alkaline granitoids into tectonically-related groups using a range of field, petrographic and geochemical criteria. In some areas it has proved possible to distinguish, using field criteria, between *I-type* (igneous source) biotite — or hornblende-biotite granitoids and *S-type* (sedimentary source) muscovite-biotite, or two-mica, granites. Although some granitoids have transitional characteristics, and others may be difficult to designate because of hydrothermal alteration, the information in Table 8.1 provides a basis for preliminary classification in the field.

The importance of calc-alkaline granitoids in continental margin arcs (e.g.

those of the circum-Pacific arc) implies a magma-genetic link with subduction processes at destructive plate margins. About 100,000 km^2 of imperfectly exposed Mesozoic and Tertiary batholith forms the mountain chains of the western Americas from Alaska to south Chile. Although much of this batholith is unstudied, detailed fieldwork in some areas has revealed considerable internal variation. Mapping of the Peru coastal batholith based on careful tracing of internal intrusive–intrusive rock contacts within the batholith over large distances, in this case made possible in the vertical dimension by the deeply-dissected nature of the Andes (Fig. 8.5), has led

Table 8.1 Field characteristics and possible origin of I- and S-type granitoids

	I-type	S-type
Overall petrology	Part of a broad spectrum from diorite or tonalite through granodiorite to granite	Restricted to leucocratic granites
Distinctive minerals	Hornblende, biotite, magnetite, sphene	Muscovite, biotite, cordierite, monazite, garnet, ilmenite
Xenoliths	Mainly igneous rocks	Mainly metasedimentary rocks
Associated metalliferous mineral deposits (see Section 8.3)	Porphyry copper and molybdenite deposits; lead-zinc deposits at lower temperatures	Tin-tungsten deposits
Origin	Produced by partial melting of igneous material	Produced by partial melting of meta-sedimentary material
Modern tectonic setting (cf. Table 1.1)	Island arcs and continental margin subduction — related batholiths	Continental collision zones with overthrust terrain

to the map shown in Fig. 8.6. This is a geological map of 120 × 70 km area showing that four major ring complexes (cf. Section 5.2.1), comprising granites, tonalites, diorites and gabbros, have cut through the main plutons of the batholith, which are dominated by tonalite. At high altitudes, the roof contacts of the batholith with overlying volcanics have been studied, whilst, at lower elevations, the different intrusive phases of the batholith have been mapped in detail, revealing a complex internal structure.

At high levels in such batholiths, individual plutons are sometimes separated by thin *screens* (average 10–100 m thickness) of metamorphic rocks, early mafic igneous rocks, or late aplite dykes. Relative ages in

igneous terraines of this type are determined using truncated structures at contacts, xenoliths of an older rock in a younger one, and dykes or net veins of a younger rock in an older. The overall composition of different plutons varies from diorite to leucogranite and, although abrupt changes in rock type are usually confined to the often knife-sharp contacts between plutons, there are also less prominent variations, particularly within plutons. For example, internal zonation from margins of intermediate composition to acid cores (Fig. 8.7) may result from *fractional crystallization* processes whereby, during crystallization, early-formed high-temperature minerals (e.g. Ca-rich feldspars, Mg-rich pyroxenes or amphiboles) accumulated in the solidifying walls of the intrusion, leaving residual liquids towards the centre of the intrusion enriched in silica and alkali elements. In this way, a primary granodiorite magma, for example, may fractionate to form a range of rock-types, from diorite and tonalite intrusion margins to granite at the centre, and the variation in rock-types across such intrusions may be completely gradational. However, this degree of zonation, derived by fractionation of a *single* magma, is unusual in a relatively small pluton (cf. Fig. 8.7) and, if variations from diorite to granite are found on this scale, then a history of *multiple intrusion* might be suspected. Clearly, *interpretation of late-stage magmatic processes depends on careful field mapping of the interiors of intrusions*, for if an intrusion of diorite were invaded by a younger granite magma, then the best way of reaching this interpretation in the field would be by finding internal intrusive contacts. Such contacts might well be complex, if the earlier intrusion was incompletely crystallized, for *magma mixing* or *hybridization* across a broad (perhaps hundreds of metres) contact zone would then be observed. Crystallized hybrid magmas may contain early-formed crystals derived from two compositionally-different magmas, each of which may be available for individual study on either side of the hybrid zone. If the earlier intrusion was solid but still warm enough to deform plastically, then diffuse contact zones, often with interpenetrating or lobate structures, would be found. Also, xenoliths of earlier crystallized rocks within the later intrusion may provide clues about the state of the earlier magma when the later intrusion was emplaced. These alternative inter-

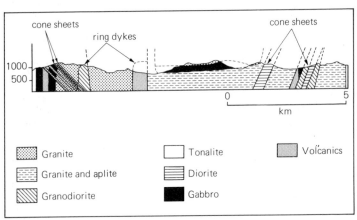

Fig. 8.5 Cross-section through the Huaura centre of central Peru showing how the relationships between the different intrusive phases, including cone-sheets, ring-dykes and larger plutons as well as the overlying volcanics, have been mapped by taking advantage of the good vertical sections exposed in the Andes.

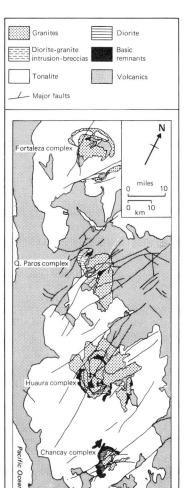

Fortaleza complex

N

miles
0   10

0   10
km

Q. Paros complex

Huaura complex

Chancay complex

Pacific Ocean

Fig. 8.6 Geological map of part of the coastal batholith in central Peru showing the distribution of four major, possibly sub-volcanic, ring complexes. The cross-section in Fig. 8.5 shows the Huaura complex in greater detail.

pretations of compositional changes may be difficult to distinguish in the field and, in many cases, may only be resolved by laboratory work.

Occasionally, igneous lamination similar to that associated with mafic–ultramafic magmas (e.g. Fig. 3.13), but involving different minerals, is observed in calc-alkaline plutons. However, layering may be much less distinct in calc-alkaline plutons because layer margins are often more gradational and mineralogical variations between the layers are usually more subtle. For example, a clear distinction between layers of tonalite alternating with granodiorite may be difficult in the field, especially if the layers are wide (say, 2–3 m) and their margins are diffuse. Occasionally, however, small-scale lamination (2–3 cm band width) may be observed, in which the mineralogical distinction between more-mafic and more-felsic bands is clear cut. Disruption of early-formed mafic layers while the felsic magma remains mobile, perhaps as a crystal mush, produces rounded mafic *autoliths* or elongate *schlieren*. The development of crystal mushes during incipient crystallization of magmas may be associated with the squeezing out of residual liquid from the consolidating mush towards the centre of the intrusion. The effectiveness of this process, known as *filter pressing*, will determine the extent to which an intrusion shows igneous lamination on a local scale or zonation on a large scale.

Most of the characteristics of calc-alkaline intrusions may be recognized in granitoid suites of all geological ages but, the older the intrusion, the greater is the probability that more complicated fabrics will have developed due to the additional effects of *metamorph-*

ism (see also Chapter 12). In such cases, any metamorphic effects, such as regional jointing or new mineral growth in bands (*foliation* which, at the scale of outcrops, may be difficult to distinguish from planar magmatic layering), should be traceable laterally into the country rocks. Granitoids which contain internal structures that have survived a later stage of metamorphism, during which new country rock fabrics, developed *around* the intrusion, are described as *pre-tectonic* (e.g. Fig. 8.8a). *Syn-tectonic* intrusions (Fig. 8.8b), including pre-tectonic intrusions from which all traces of an earlier fabric are lost and intrusions emplaced during metamorphism, tend to have internal flow structures parallel to the fabrics in adjacent country rock schists or gneisses (i.e. foliated or compositionally banded rocks, respectively; see Fry, 1984). Some syn-tectonic intrusions are finely interlaminated (on a centimetre to metre scale) with metamorphic rocks and are then termed *migmatites* or, if there are also small discordant igneous veins, *injection complexes*. The terms migmatite is used to describe rocks that appear to have been emplaced as a *migma*, which is a mobile mixture of pre-existing solid material together with some partial melt (further details in Chapter 12). In contrast, unfoliated intrusions with discordant margins, which may be emplaced into previously metamorphosed rocks at high crus-

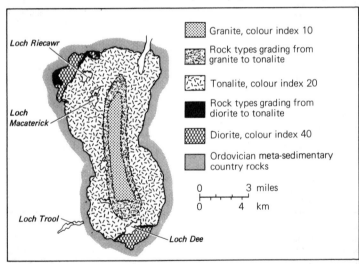

Fig. 8.7 Example of a zoned granitoid pluton (Loch Doon) from the Southern Uplands of Scotland. In the field there is a fairly rapid transition from the marginal diorites, which may represent an early separate intrusive phase, to tonalite over about 100 m, though knife-sharp contacts are rare. In contrast, there is a complete gradation from tonalite to granite throughout the remainder of the intrusion; the 1000 m wide boundary zone was mapped using the type and amount (colour index) of mafic minerals at outcrop.

tal levels, are *post-tectonic* (Fig. 8.8c). The block diagram in Fig. 8.8d shows a region in which the shape and metamorphic fabrics of syn-tectonic bodies are easily distinguished from those of a post-tectonic intrusion.

Intrusions with different relationships to regional metamorphism may sometimes occur at the same level of exposure (Fig. 8.8d); for example, Fig. 8.9 illustrates a typical deeply-eroded area of southwest Finland where a syn-tectonic calc-alkaline suite (dated at 1900 Ma), with east–west flow structures parallel to the foliation in local amphibolitic gneisses, is cut by a later post-tectonic (1600 Ma) alkali granite. The calc-alkaline rocks are dominated by *trondhjemite* with lesser amounts of tonalite and gabbro. (Trondhjemite is a name used for leucotonalites (cf. Figs 4.8, 4.9) dominated by

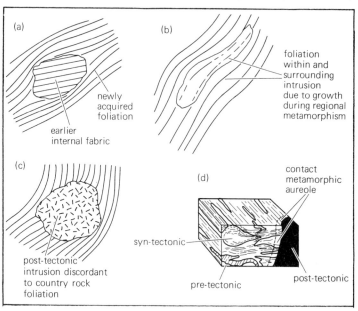

Fig. 8.8 The relationship between intrusive rocks and phases of deformation: (a) A pre-tectonic intrusion with an earlier internal fabric, around which metamorphic fabrics have been developed without affecting the intrusion. (b) Syn-tectonic intrusion emplaced during regional metamorphism, and controlled in terms of both shape and internal fabric by the tectonic stress regime. (c) Post-tectonic intrusion emplaced through the discordant to pre-existing fabrics. (d) Block diagram showing syn-tectonic and pre-tectonic intrusions (light shading) developed due to strong vertical and moderate left-to-right compression in three dimensions. Fine lines indicate mineral lineations in the syn-tectonic intrusion and host rocks. Solid ornament (right) denotes a post-tectonic intrusion.

quartz and sodium-rich plagioclase feldspar in which alkali feldspar is less than 10% of the total feldspar and the colour index is less than 10.) This raises a general point about long-term changes in the petrology of calc-alkaline intrusions; whereas modern batholiths are composed of a granite-granodiorite-diorite or tonalite-gabbro suite, in older examples from the early Proterozoic and Archaean (i.e. > 1500 Ma before present) a trondhjemite-tonalite-gabbro suite is more common. So, in the field, ancient and modern calc-alkaline intrusions may contrast in both their degree of metamorphism and in their petrology. However, the contrast is also a function of tectonic setting and erosion level, because the plutons of young island arcs and the

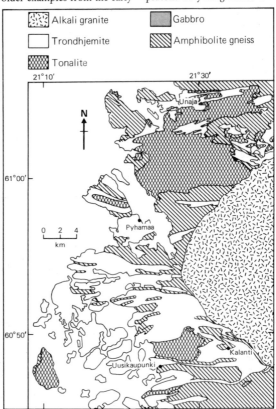

Fig. 8.9 Simplified geological map of southwest Finland showing syn-tectonic calc-alkaline tonalite and amphibolitic gneisses in the west intruded by a post-tectonic alkali (rapakivi — see Section 8.4) granite in the east.

108

more deeply-eroded parts of the Andean batholith both contain more trondhjemitic rocks and less granitic rocks.

8.3 Metalliferous mineralization

Mineralization associated with igneous rocks, and calc-alkaline suites in particular, is related to the emplacement of a hot magma body at a high level in the crust which may initiate hydrothermal convection within both the intrusion and the surrounding permeable country rocks (see discussion of hydrothermal veins in Section 3.2). In some cases radiothermal heat associated with K,U,Th-rich granites may stimulate hydrothermal convection of meteoric groundwaters (i.e. those derived by downwards-seeping rainwater) long after the intrusion has solidified. Trace elements — particularly base metals (iron, copper, lead, zinc), tin and tungsten, derived from the igneous body and/or from leaching of the country rocks by hydrothermal fluids — may be concentrated and precipitated when the fluids cool as disseminated oxide or sulphide crystals in the igneous rock or as veins, exploiting a joint network (e.g. Fig. 3.5) that may also invade the country rock. Disseminated mineralization, usually most prominant near the periphery of a porphyritic diorite or porphyritic granodiorite intrusion, is termed a *porphyry mineral deposit* (notably the massive porphyry copper deposits which are dominated by chalcopyrite and pyrite), and networks of small veins, again usually concentrated near intrusion–country rock contact zones, are known as *stockworks* (see examples in Fig. 8.10). Veins usually contain one or more minerals which are abundant but do not have commercial value, and are termed *gangue* minerals (e.g. quartz, calcite, barytes) together with a lesser quantity (perhaps a few per cent) of metalliferous minerals; they may be simple (one generation of minerals) or complex (two or more generations; Fig. 3.4). Table 8.2 lists the properties of some of the more common metalliferous mineral grains that might be found in association with calc-alkaline igneous intrusions. Careful examination of the Table shows that most of these metallic minerals may be distinguished from one another using colour and hardness criteria. *The nature of the vein infillings — metalliferous and gangue crystals and their proportions, simple or complex veins — together with their width, orientation and frequency should be studied in the field; it is also important to note the characteristics of the wall rocks*, as explained below.

Associated with mineralization, the host rocks become altered by hydrothermal fluids, the effects of which are usually more pervasive and extensive than the mineralization itself. Therefore, field occurrences of hydrothermal alteration may provide a means of determining whether the region around or within an igneous body might be mineralized. Common forms that may be encountered and recognized in field samples are (cf. Table 4.6):

(i) *silicification*, in which quartz overgrowths and replacements of existing minerals occurs;

(ii) *kaolinization*, or *argillic alteration* — the alteration of feldspar to clay minerals;

(iii) *greisenization* — the alteration to alkali feldspar to secondary white mica, sericite;

(iv) *propylitic alteration*, in which plagioclase feldspar alters to epidote

(a)

(b)

Fig. 8.10 (a) Stockwork of mineralized veinlets in porphyritic granite from the El Salvador porphyry copper deposit, N. Chile. (b) Vein-type mineralization showing the edge of a wolfram-quartz vein cutting hornfelsed Skiddaw Slates, Carrock Mine, Cumbria. The edge of the vein is at the bottom right of this picture. Sample in (a) is 10 × 6 cm; field of view in (b) is 1 × 0.6 m.

and mafic minerals to chlorite;

(v) *potassium silicate alteration*, in which plagioclase feldspar and amphibole alter to K-feldspar and biotite mica.

For example, an outcrop of greisenized granite with disseminated pyrite (FeS_2) (e.g. Fig. 3.5) may indicate that deposits of chalcopyrite ($CuFeS_2$) are not far away. Fig. 8.11 illustrates, in cross-section, the way in which both wall-rock alteration (e.g. from K-silicate to greisenization to kaolinization or propylitic alteration) and the nature of the metalliferous minerals precipitated (e.g. from chalcopyrite to pyrite for Cu-Fe bearing veins) both change with falling fluid temperatures and with distance from the centre of a mineralized stockwork. Thus, concentric zones of alteration and mineralization may be recognized in field studies.

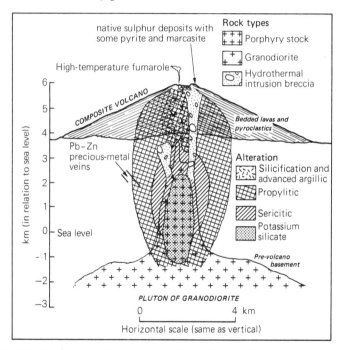

Fig. 8.11 Summary of typical zones of wall-rock alteration that occur in and around porphyry copper mineral deposits.

Table 8.2 Characteristics of common metalliferous minerals found in association

Mineral	Chemical formula	Usual colour and streak
Chalcopyrite	$CuFeS_2$	Brass yellow to golden yellow, sometimes irridescent tarnish; black streak
Pyrite	FeS_2	Pale brass yellow to bronze yellow; greenish-black streak
Arsenopyrite	$FeAsS$	Silver grey to steel grey; black streak
Molybdenite	MoS_2	Lead grey, occasionally greenish-grey; dark grey streak
Galena	PbS	Lead grey, sometimes tarnishes; dark grey streak
Sphalerite	ZnS	Yellow-brown, good crystals sometimes black; yellow-brown streak
Wolfram	$FeWO_4$	Black, sometimes dark brown often associated with scheelite; dark yellow-brown streak
Scheelite	$CaWO_4$	White, pale green or pale yellow; white streak
Cassiterite	SnO_2	White to pale yellow; pale yellow streak

with igneous intrusions

Lustre	Usual form	Hardness	Density ($\times 10^{-3}$ kg m^{-3})
Metallic	Usually occurs as vein encrustations, occasional small cubic crystals	3.5–4.0	4.1–4.3
Metallic	Smooth, shiny or striated crystals, often forms good cubes	6.0–6.5	5.0–5.2
Metallic, steely	Massive, occasionally radiating clusters	5.5–6.0	5.9–6.2
Strongly metallic	Massive, scaly aggregates of flexible laminae. (NB Very soft, scratched by finger nail)	1.0–1.5	4.7–4.8
Metallic	Massive cubic crystals, but occasional fibrous crystals	2.5	7.2–7.6
Resinous	Massive, but occasional fibrous or cubic crystals	3.5–4.0	3.9–4.2
Sub-metallic	Tabular, often forms striated, bladed crystals	5.0–5.5	7.1–7.5
Vitreous to resinous (cf. feldspar, but fluoresces in UV light)	Usually massive, occasional rectangular prisms, sometimes platy and striated	4.5–5.0	5.9–6.1
Brilliant on crystal faces, otherwise dull	Massive, fibrous, usually disseminated small grains	6.0–7.0	6.8–7.1

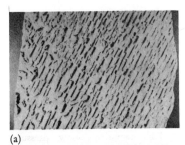

(a)

(b)

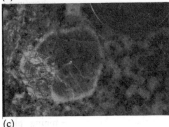

(c)

Fig. 8.12 (a) *Graphic* granite containing quartz crystals in a feldspar matrix. The crystals sometimes look like ancient hieroglyphic (or runic) characters. Area shown is 7 × 10 cm. (b) *Orbicular* granite containing orbicules set in a granitic groudmass. Individual parts of the orbicules vary in their relative proportions of feldspar and mafic minerals. Area shown is 7 × 10 cm. (c) Feldspar phenocryst in Shap granite (Cumbria, England) showing *rapakivi* texture, involving the growth of plagioclase mantles around alkali feldspar cores. Area shown is 3 × 5 cm.

8.4 Distinctive granitoid textures

Here are brief descriptions of several distinctive, though rather uncommon, textural varieties that may be encountered in the field, among granitoids of various types.

Graphic granite refers to an intergrowth texture which develops in some quartz-alkali feldspar pegmatites. Aligned, rod-like, or triangular light grey blebs of quartz are enclosed within large feldspar crystals (Fig. 8.12a) and take on the appearance of runic characters. If the quartz-alkali feldspar intergrowths are confined to smaller crystals, perhaps in the groundmass of a porphyritic granitoid, then the texture is *granophyric* (and is often only visible in thin section). Rocks with granophyric texture are usually fine-grained porphyritic granodiorites and granites ('high-level' granites, Section 7.3) and they occur most frequently as minor intrusions.

Orbicular granite (Fig. 8.12b) contains large ovoid bodies (orbicules), usually up to several centimetres in diameter, in which the components are arranged in concentric layers of leucocratic quartz — feldspar-rich material alternating with mafic layers consisting primarily of dark hornblende and/or biotite. Each orbicule contains a core, commonly an igneous xenolith fragment but sometimes composed of the same granitic material in which the orbs are embedded. The thicknesses and mineral contents of the layers forming different shells of the orbicules are often similar, suggesting that they grew together. This may have happened during rhythmic crystallization, analogous to that responsible for igneous lamination (Sections 3.3 and 8.2), with the difference that, instead of resulting from crystal settling, the orbicules accreted available crystals from the magma before final consolidation of leucocratic material in the interstices took place.

Rapakivi granite (Fig. 8.12c) contains large phenocrysts (several centimetres long) of alkali feldspar, often salmon pink or flesh-coloured, mantled by a rim (1–2 mm thick) of white plagioclase feldspar. These complex phenocrysts are usually embedded in a groundmass of quartz, feldspars, biotite and/or amphibole. Often the alkali feldspars are rounded, unlike most phenocrysts in igneous rocks which are euhedral, possibly indicating corrosion or remelting of their edges into the magma before crystallization of the plagioclase rim. Rapakivi texture is most common in alkali granites (Chapter 9), notably the type localities in south Finland, but is also found in some calc-alkaline types.

8.5 Metamorphic aureoles

This brief section on metamorphic aureoles is included here because aureoles are best developed around granitoid (granodiorite and granite) intrusions emplaced within a few kilometres of the Earth's surface, where there is liable to be a marked contrast between the temperature of the intrusion (*ca.*700–1000°C) and the country rocks (< 200°C). Aureoles reflecting metamorphism in the temperature range 200–700°C are characteristically up to 2–3 km wide around granitoid intrusions 10–15 km in diameter, where the margins dip at 60–70° below the surface (Fig. 8.13). Generally, the aureoles of basic intrusions such as dykes and sills are much smaller; for example, many large sills with thicknesses of up to several hundred metres may show metamorphic effects up to only a few metres from the contact (see also Section 5.2). The textures and mineralogical changes developed by contact metamorphsm of common rock-types are summarized in Table 8.3; these characteristic minerals provide a basis for mapping and studying aureole zones such as those discussed for the high-level granites in Secion 7.3. For further details of high temperature–low pressure metamorphism of sedimentary sequences, refer to Fry (1984).

8.6 Summary of the field characteristics of calc-alkaline intrusions

1 Calc-alkaline plutons comprise different proportions of coarse-grained rocks: granite, granodiorite and tonalite, with subordinate diorite and gabbro. Plutons may show evidence of emplacement by stoping; that is, the fracturing of the roof and incorporation of country rock xenoliths. Together, a large number of individual plutons (each 1–50 km in cross-section at outcrop, and roughly equidimensional), may comprise a batholith. When examining plutons, it is necessary to:

   (i) Describe and determine the relative proportions of rock-types present; use Table 8.1 if a preliminary tectonic interpretation is required.

   (ii) Examine contact relations to determine age relations.

   (iii) Study the nature and distribution of xenoliths.

2 Examination of individual plutons may reveal small-scale layering or lamination (0.01–10 m), or zoning

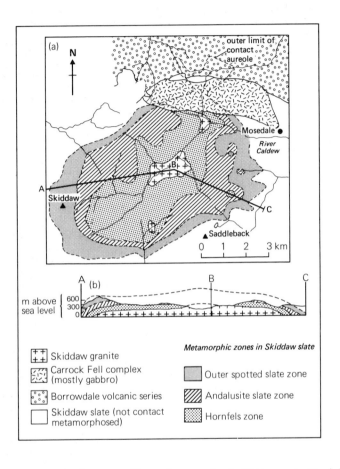

Fig. 8.13 Example of a metamorphic aureole surrounding the Skiddaw granite (Cumbria, England), (a) in plan view and (b) in cross-section. The inner zone of metamorphism (coarse stipple) surrounding the three granite outcrops is characterized by crystalline hornfels (produced at temperatures greater than about 500°C, see Fig. 8.14b) whereas the outer zone (fine stipple) contains mainly spotted slates (produced the lower temperatures, e.g. Fig. 8.14a). Both spotted slate and hornfels are typical of contact aureoles developed in pelitic (shaley) country rocks (Table 8.3). Notice that mapping of the contract aureole gives a good impression of the sub-surface intrusion shape and implies a shallow depth to granite between the three known outcrops.

Table 8.3 Mineralogical and textural effects of contact metamorphism of common rock-types (for details of minerals and textures not defined here, refer to Fry, 1984.)

Pure limestone	No mineralogical changes; increase in degree of recrystallization as contact is approached, leading to formation of completely recrystallized *marble*. The thickness of the zone of contact metamorphism and recrystallization varies from a few centimetres adjacent to minor intrusions, to hundreds of metres adjacent to larger plutonic intrusions.
Impure limestone (inc. dolomite)	Complex sequence of mineralogical changes; the following mineralogical assemblages may develop as the contact is approached: (i) calcite ± dolomite ± clay minerals; (ii) calcite ± dolomite ± olivine (Mg-rich olivine, forsterite) ± amphibole (tremolite); (iii) calcite ± serpentine ± chlorite ± tremolite; (iv) wollastonite (Ca-pyroxene) ± garnet (Ca-Mg-Fe rich garnet, grossularite and andradite) ± pyroxene (Ca-Mg rich pyroxene diopside). These changes are accompanied by recrystallization to marble. The thickness of the zones corresponds to those described for pure limestone above. Typically, for a metamorphic aureole several hundred metres wide, the zones with diopside and wollastonite would be only a few metres thick.
Shale	Complex sequence of mineralogical changes; the following mineralogical changes may develop as the contact is approached: (i) development of 'spots'; (ii) large crystals (or porphyroblasts) of andalusite (chiastolite) (Fig. 8.14a); (iii) cordierite-andalusite-biotite. These changes are accompanied by progressive recrystallization to hornfels-textured rocks (these are usually banded and frequently folded on a fine scale, Fig. 8.14b). These zones may all be developed within 1–2 km of the igneous contact (cf. Fig. 8.13).
Sandstones	No mineralogical changes; recrystallization limited to within a few metres of the contact, and does not usually exceed a few tens of metres even around large intrusions.
Basic volcanic rocks	Complex sequence of mineralogical changes; the following mineralogical assemblages may develop as the contact is approached: (i) epidote-chlorite-amphibole; (ii) plagioclase-hornblende; (iii) plagioclase-pyroxene. These changes are accompanied by progressive recrystallization to form basic hornfels-textured rocks (as above) within 1–2 km of the igneous contact.

on a pluton scale. These may be due to combinations of crystal accumulation and filter pressing. Plutons may also contain igneous rocks derived by multiple injections of different magmas between which contacts may be recognized, with associated igneous xenoliths. Important activities, therefore, are to:

(i) Describe the nature and orientation of any igneous lamination.

(ii) Decide whether a pluton is zoned on a regional scale and look for internal contacts.

(iii) Study the nature and distribution of autoliths.

3 Other internal structures of plutons may include linear or planar fabrics which may be interpreted in terms of magmatic flow or of post-magmatic deformation. The relationship of such fabrics to those in surrounding metamorphic rocks should be described using the pre-, syn- and post-tectonic nomenclature shown in Fig. 8.8.

4 Disseminated or vein type (stockwork) metalliferous mineralization with 'haloes' of characteristic wall-rock alteration, both due to hydrothermal activity, is common near the boundaries of high-level plutons intruding fractured, permeable (therefore, low-grade) metamorphic or sedimentary strata. In such cases, it is necessary to:

(i) Describe the relationship of mineralization to the host intrusion.

(ii) Identify the metalliferous mineral present using Table 8.2.

5 Distinctive granitoid textures — graphic granite, orbicular granite and rapakivi granite — should be described with the aid of Fig. 8.12.

6 The type of contact metamorphism in the country rocks surrounding high-level granites may be distinguished with the aid of Table 8.3.

(a)

(b)

Fig. 8.14 (a) Elongate crystals of andalusite (chiastolite) in slightly hornfelsed slate. Field of view is 4 × 3 cm. (b) Strongly hornfelsed slate (cordierite-andalusite hornfels) showing development of fine banding in a zone immediately surrounding the Skiddaw granite (dense stipple in Fig. 8.13). Length of hammer is 25 cm.

9
Plutonic rocks II:
the alkaline association

9.1 Definitions, general features and occurrence

The term 'alkaline association' is applied to both volcanic and intrusive associations varying from basic through intermediate to acid compositions in terms of their silica saturation, but which are rich in the alkali elements, sodium and potassium. Alkaline associations are usually defined in purely chemical terms, using plots of K_2O or $Na_2O + K_2O$ against SiO_2 resulting from laboratory analysis. Such definitions are reasonably consistent with mineralogical distinctions that can be recognized in the field, such as the abundance of *alkali feldspar* in many mesocratic and leucocratic rocks, and the occurrence of *foid minerals* in mafic (e.g. alkali basalt or gabbro) and mesocratic (e.g. nepheline syenite) rocks. Granites and syenites are sometimes termed alkaline on the basis of having a very high ratio (> 9 : 1) of alkali to plagioclase feldspar — this mineralogical definition is embodied in the QAPF classification (cf. Fig. 4.8 and Appendix I) which is recommended for field use (see below). *Independently*, for intermediate and acid volcanic and intrusive rocks, the *molecular* balance of alkali elements, calcium and aluminium is often used to distinguish *peralkaline* (molar $Na_2O + K_2O$ > molar Al_2O_3) rocks, which commonly contain blue or green alkali pyroxenes and amphiboles. Distinctive alkaline granites are termed *A-type* in some recent literature to distinguish them from I- and S-types (Table 8.1).

There are two kinds of acid-basic 'alkaline associations' that may be defined using the QAPF diagram: the alkali feldspar granite–granite–syenite–gabbro association (labelled '1' in Fig. 9.1a) which is often bimodal, (i.e. granite and gabbro with no syenite) and the alkali feldspar syenite–syenite–foid syenite–foid gabbro association (labelled '2' in Fig. 9.1a). In the field it will usually be possible to recognize which of these associations is represented, particularly from the most acid rocks which will be rich in alkali feldspar and have more (alkali feldspar granite) or less (alkali feldspar syenite) than 20% visible quartz. Accompanying either kind of alkali association, and possibly genetically related to the alkaline rocks as the most basic rocks present, massive sheet-like anorthosite bodies sometimes occur (see Chapter 11 for further details).

Alkaline rocks are very much less common than calc-alkaline types, with the possible exception of parts of the

Precambrian geological record. Most of the better known examples, often with abundant rapakivi granites (Section 8.4), are confined to Proterozoic crust, where they occur commonly in extensive fault-bounded blocks ranging for hundreds of kilometres in length where the lithosphere was under tension. But there are many younger examples, such as the granites of igneous centres that developed within the Scottish Tertiary Volcanic Province. A typical centre might include several cross-cutting high-level plutons (e.g. Fig. 9.1b) together with ring-dykes, cone-sheets, linear dyke swarms and lavas (Chapters 6 and 7) and, to obtain a proper understanding of magmatic evolution within a centre, the age relations of these different components should be established in the field. Although the size of each intrusion, ranging from 1 to 20 km in diameter, sometimes reaches that of individual plutons within calc-alkaline batholiths, the relatively young intrusions belonging to alkaline rock associations are generally confined to individual centres, or groups of centres, rather than forming extensive continuous batholiths.

Metalliferous mineralization is associated with both magmatic and metasomatic alkaline complexes. It may include lead-zinc deposits, but cassiterite and wolfram (Table 8.2) are typically abundant, and often occur with other uncommon minerals, such as green *beryl* ($Be_3Al_2Si_6O_{18}$), black *uraninite* (which, when massive, is known as pitchblende — UO_2-U_3O_8), purple *fluorite* (CaF_2) and silver-grey *lepidolite* (lithium mica —KLi_2 $(AlSi_3O_{10})$ $(OH)_2$).

Apart from the points of contrast already noted, in most other ways alkaline and calc-alkaline intrusives

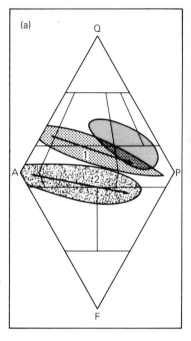

Fig. 9.1 (a) Simplified QAPF diagram showing data for Tertiary alkali feldspar granites and granites from the Isle of Skye, Scotland — trend 1; the field of alkali feldspar syenite–syenite–foid syenite–foid gabbro suites (such as the Caledonian alkaline intrusive suites of northwest Scotland) — trend 2; and the field (grey tone) of calc-alkaline rock suites. Note that the different alkaline rock series vary in their quartz content such that their trends (lines 1 and 2) are usually almost parallel, starting at the QAF side of this diagram and moving towards the QPF side. Alkali rock series almost always include some alkali feldspar granites or alkali feldspar syenites — a feature that usefully distinguishes between them and calc-alkaline series in the field.

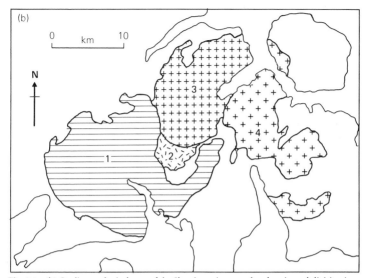

Fig. 9.1 (b) Outline geological map of the Skye intrusive complex showing subdivision into centres which, from evidence of intrusive contacts, developed in the following time sequence: 1, The Cuillin Hills; 2, Strath na Creitheach; 3, Western Red Hills; 4, Eastern Red Hills. Centre 1 is composed mainly of mafic-ultramafic rocks; centres 2–4 are mainly alkali feldspar granites and granites.

have similar features, such that alkaline complexes frequently display internal contacts due to multiple intrusion (e.g. Fig. 9.2); they may show zonation or igneous lamination and, since they are usually emplaced at a high level and are post-tectonic (to the most recent regional metamorphism), they often have prominent contact metamorphic aureoles.

9.2 Distinctive alkaline rock-types and their mineralogy

Alkaline rocks are often highly distinctive in appearance and the following notes summarize some of their major

petrological characteristics (see also Table 9.1).

9.2.1 Alkali feldspar granites

These are often very leucocratic coarse- to medium-grained rocks with abundant quartz but in which it may be difficult to recognize the type of feldspar present. However, by definition, alkali feldspar exceeds 90% of the feldspars present in an alkali feldspar granite. Tiny blue-grey needles of alkali amphibole riebeckite, or rectangular crystals of green pyroxene, aegirine (or red-brown acmite), perhaps forming up to 15% of the rock, are characteristic minor consti-

121

tuents of peralkaline alkali feldspar granites and granites. Other peraluminous and metaluminous granites may still be alkaline in the QAPF diagram if they contain abundant alkali feldspar with either two micas (see Table 8.1)

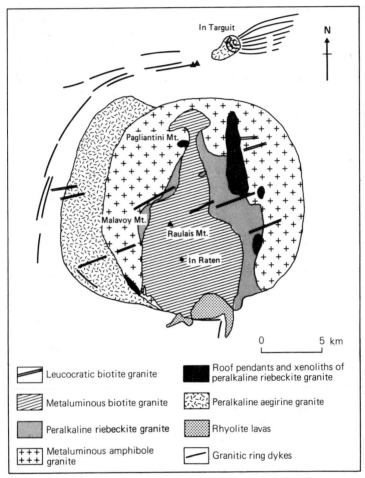

Fig. 9.2 The Tarraouadji massif of the Niger Republic, one of the Jurassic granite complexes of the Niger–Nigerian region. The centre includes metaluminous and peralkaline granite types in order of increasing age down the key and from left to right (note that the rock types are mainly granites with a few alkali granites in the QAPF diagram).

or with amphibole and biotite mica (Table 9.1). Figure 9.2 shows that, in a single alkaline centre, it is common for a spectrum of metaluminous and peralkaline granites to occur in close association.

9.2.2 Alkali feldspar syenites and syenites

These rocks contain less than 20% quartz and their most distinctive feature is the presence of coarse white to salmon-coloured grains of alkali feldspar (up to 80% of the rock) together with 0–35% (typically 20%) of mafic minerals. Alkali feldspar syenites have more than 90% of their feldspars as alkali feldspar whereas syenites have 35–90% of their feldspar as alkali feldspar (Table 9.1). Metaluminous alkali feldspar syenites and syenites contain dark, equidimensional or rectangular prismatic crystals of amphibole, sometimes with biotite mica, (see Table 9.1). Peralkaline alkali feldspar syenites and syenites may have the same range of mafic minerals as for peralkaline alkali feldspar granites and granites (above), and the two groups of rock are distinguished from each other by quartz content. As noted earlier, it is unusual to find alkali feldspar syenites and alkali feldspar granites occurring together.

9.2.3 Foid syenites and gabbros

Nepheline-bearing syenites are the most common types and, since nepheline is often grey or greasy in appearance (Table 4.6), it may be easily

Table 9.1 Mineralogical compositions of common rocks occurring in alkaline associations (cf. Fig. 4.8)

Quartz or foids % of whole rock	Alkali feldspar as % of total feldspar			
	> 90	50–90	35–50	< 35
Quartz > 20	Alkali feldspar granite	Granite		Granodiorite (unusual in alkaline suites)
Quartz < 20% Foids < 10%	Alkali feldspar syenite	Syenite		Gabbro or diorite
Foids > 10%	Foid-bearing syenite		Foid gabbro or diorite	

The different groups of granites and syenites, as defined geochemically, may be distinguished by mineralogical criteria as follows:

peralkaline — mafic minerals are alkali amphibole or pyroxene, micas are rare or absent
metaluminous — mafic mineral is calcic amphibole, (cf. Table 4.6) occasionally with biotite mica
peraluminous — mafic mineral is biotite mica; muscovite mica is sometimes also present: peraluminous rocks are least common in alkaline suites

Gabbro and diorite are distinguished from each other on the basis of feldspar composition (cf. Section 4.5), though a widely applicable field criterion is that the former contains pyroxenes, and the latter biotite and amphibole.

misidentified as quartz. The main problem, therefore, is that of identifying small amounts, perhaps only 10%, of foid minerals in hand specimens which, again, are dominated by alkali feldspars and lesser amounts of mafic minerals. The best ways of overcoming this problem in the field is to examine weathered surfaces, where quartz tends to be smooth whereas nepheline becomes pitted and etched by alteration. With increasing foid mineral content from 10 to 60% (Fig. 4.8) foid syenites become progressively easier to identify, particularly as the nepheline may take on a greenish colour and, more especially if blue or yellow foid minerals (sodalite and cancrinite) are present. Foid gabbros in alkaline associations usually have pale-grey, green or white feldspar, comprising about 50–60% of the rock. Rocks which are considerably richer than this in feldspar are often also found and

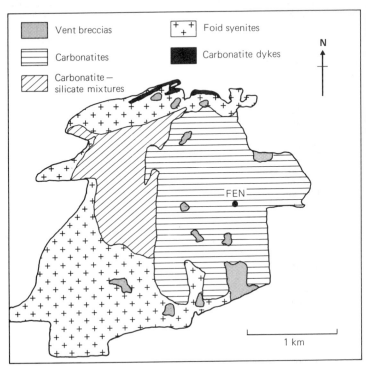

Fig. 9.3 The Fen alkaline igneous complex of South Norway. Rock types within the complex include volcanic vent breccias, carbonatites, mixed carbonatite-silicate rocks and various forms of foid syenite. Carbonatite dykes are in black and the whole complex is set within a region of ancient calc-alkaline granite gneisses.

these are better described as anortho-sites (see Fig. 4.10 and Chapter 11).

9.2.4 Fenites

Certain types of post-intrusive alte-ration may be responsible for the pro-duction of highly alkaline rocks from calc-alkaline or other igneous materials; for example, some Precam-brian examples apparently developed in association with emplacement of carbonatites (Section 4.5) and are interpreted as being formed by *alkali metasomatism*. This involves the alte-ration of calcic or calc-alkaline rocks by CO_2-rich vapours that leach calc-ium and add alkalis, thereby producing rocks of alkali granite, syenite and foid syenite composition. The Fen alkaline complex near Oslo (Fig. 9.3) is a typ-ical example where volcanic vent brec-cias cut through carbonatites and syenites. Around the vents are alkali-enriched metasomatic rocks containing alkali feldspar and alkali pyroxenes replacing the original mineralogy; such recognizably metasomatic rocks are collectively known as *fenites*, and the process by which they are formed is termed *fenitization*.

9.3 Summary of the field character-istics of alkaline intrusions

1 Suites of alkaline intrusive rocks tend to occur in groups of igneous centres characterized by plutonic, minor intrusive and sometimes extrusive activity. Individual cen-tres may lie on regional trends but they do not form continuous bath-oliths. Apart from distinctive rock-types and mineralization (Sn-W-U is most common), in all other sen-ses the field relations for alkaline intrusions resemble those for calc-alkaline intrusions (Section 8.2)

2 The main rock-types, summarized in Table 9.1, are generally poorer in mafic minerals but richer in alkali feldspar, and sometimes in foids or alkali pyroxenes and amphiboles, than calc-alkaline equivalents. They form acid-basic (sometimes bimodal) associations as indicated in Fig. 9.1.

3 Some centres which have alkaline rocks with an origin due to alkali metasomatism, collectively known as fenites, may occur in close association with carbonatites.

Plutonic rocks III: mafic–ultramafic associations

10.1 General features and occurrence

Although the mafic rock-types, gabbro (35–65% plagioclase feldspar, 35–65% pyroxene and olivine) and melagabbro (65–85% mafic minerals) are well known as minor constituents of calc-alkaline and alkaline granitoid suites (Chapters 8 and 9), they are the major constituents of mafic intrusions and are important constituents of ultramafic intrusions (where most of the rocks contain over 90% mafic minerals). These rocks occur in two main settings. First, they occur as isolated intrusions ranging from a few to hundreds of kilometres across with circular or equidimensional intrusion cross-sections. Such intrusions have been emplaced in within-plate ocean island and continental margin settings (e.g. the islands of Rhum and Skye in north-west Scotland). Second, mafic-ultramafic rocks occur as linear groups of intrusions, representing former oceanic lithosphere, which were emplaced tectonically by *obduction* (over-thrusting) at continent–continent or continent–island arc suture zones. The latter are known as *ophiolite complexes*; they include characteristic associations of lavas, dykes and mafic-ultramafic intrusions.

A wide variety of different rock-types may be present in a mafic-ultramafic intrusion. The mafic rocks vary from clinopyroxene-bearing gabbros through to rocks in which orthopyroxene is more abundant, and for which the name *norite* is often used instead of gabbro. Thus, norite is a gabbro-like rock with 35–60% of plagioclase feldspar and 35–65% of orthopyroxene, which is often bronze-coloured. However, it is the *ultramafic* rocks that show the greatest mineralogical diversity. When the mafic mineral constituents exceed 80% of the rock, then one or more of olivine, clinopyroxene and orthopyroxene may become major constituents, and this leads to the discrimination of peridotites (> 40% olivine), pyroxenites (> 60% pyroxene) and the other less common varieties shown in Fig. 4.11.

In general, *peridotite* is usually a mid-green to black, medium or coarse-grained rock-type when fresh, though the olivine in many examples has become altered to the dull, light-green colour of serpentinite (a soft, fibrous, hydrated magnesium silicate, Fig. 10.1). Serpentinization is also common in *dunite*, a medium- to coarse-grained rock almost entirely made (> 90%) of olivine which, when fresh, may have a 'sugary' texture. *Pyroxenites* have less tendency to alteration and are usually black or greenish-black in colour, though abundant orthopyroxene may have a bronze colouration. Distinctive accessory minerals, particularly in continental ultramafic bodies and kimberlite xenoliths (Section 7.2), include occasional shiny grains of *phlogopite mica* and less commonly Fe-Ti oxide mineral grains such as ilmenite, magnetite and chromite (Table 4.6) which form small black granular clusters. A rare, but economically important and easily identified rock type, *chromitite* (Fig. 10.2), occurs as layers in ultramafic intrusions. This is a heavy medium grained melanocratic granular rock in which black, euhedral chromite grains are enclosed in a matrix of clinopyroxene and/or olivine.

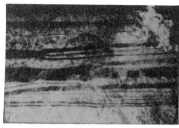

Fig. 10.2 Layered anorthosite (light) chromite (dark) outcrop in the Bushveld Complex, South Africa. Area of picture 4 × 2 m.

Fig. 10.1 Fibres of serpentinite in altered peridotite from South Norway. Width of specimen 15 cm.

10.2 Continental mafic-ultramafic intrusions

These bodies are large, discordant, funnel- or saucer-shaped intrusions, generally with inward dipping contacts, causing them to be thicker in the centre than at the perimeter, i.e. they are *lopoliths* (cf. Chapter 5). Most of them contain medium- to coarse-grained ultramafic, gabbroic, leucogabbro and even anorthositic (Chapter 11) lithologies as shown in Fig. 10.3; together these rock-types form a series of repetitive, or rhythmic layers (with igneous lamination) which crystallized within a magma chamber fed by mafic/ultramafic magma. The terminology for layered rocks is based on classic studies which formulated the concept that the layers represent concentrations of minerals accumulated by mineral settling; the layered rocks have been termed *cumulates*. However, it is now realized that some layered rocks cannot be a result of such crystal settling, and that layering might form by a number of processes. Hence the term 'cumulate' may be applied to such rocks in a non-genetic sense, in which crystal settling is a possible but non-essential process in the origin of cumulate rocks.

Layers and laminae (see Section 3.3 for definitions) vary greatly in composition, texture and lateral extent. They may be *graded* in terms of mineral content and grain-size, and the commonest case is for upward-grading from dense to less-dense (dark- to light-coloured) minerals, or from coarse- to fine-grain size (Fig. 10.4).

The contacts between layers vary between sharp and gradational, and between concordant and discordant with the adjacent layer. Both layers and laminae may be qualified by the terms 'grain-size', 'modal' (i.e. mineral proportion) and 'textural' for a field description. Hence it is possible to describe an outcrop of gabbroic cumulate as being characterized by 'grain-size and modal layering in which there is upward-gradation from melagabbro with large pyroxene crystals to leucogabbro with smaller pyroxene crystals, showing planar lamination of plagioclase' (cf. Fig. 3.13).

Fig. 10.4 Graded layering in gabbros from the Bushveld Complex, South Africa. Lens cap is 5 cm in diameter.

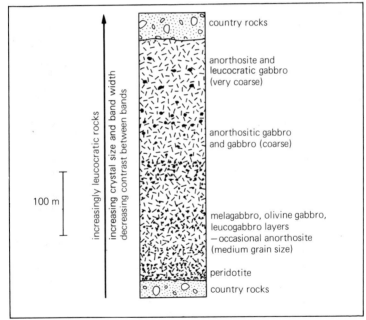

Fig. 10.3 Schematic vertical section through a layered mafic-ultramafic intrusion showing various features described in the text. Black spots and dashes indicate, respectively, the relative proportions of mafic and felsic minerals; the sizes of the symbols give an impression of changing grain size. There would be many more layers than shown here.

In many intrusions, the layering may show complex medium-scale structures, such as channelling, cross-stratification (of discordant layers), slumping and complex convolution within layering (e.g. Figs 3.14, 3.15). In some cases, xenoliths of early-crystallized cumulate appear to have slumped into younger layered cumulates. These features are similar to common sedimentary structures and their occurrence provides evidence of processes that deform and slump soft sediments and do *not* necessarily require crystal settling.

In describing layered rocks, large-scale repetition of particular types of layering may be observed. Such repetition is usually conspicuous, involving the systematic recurrence of distinctive layers or sequences of the same kind of layers, and is termed *rhythmic layering* (e.g. uppermost and lowermost layers in Fig. 3.13). These were originally described from the Skaergaard intrusion, which is characterized by the occurrence of modally graded units, 5–50 cm in thickness. Finally, many cumulate rocks are also associated with coarse-grained segregations, such as *pegmatite* (sharp margins) and *schlieren* (diffuse margins), and later minor intrusions, such as dykes and sills.

On a broad scale, in most of the large-scale intrusions, the proportion of felsic material increases upwards (e.g. Fig. 10.3), in some cases giving way to granitic rocks in the uppermost layers. Individual layers vary upwards from a few centimetres (most typical) to metres and perhaps tens of metres (exceptionally in the upper regions of an intrusion where the contrast between layers tends to be much less than near the base). *Accurate observations of band orientations, widths and their variation normal to the layering therefore provide important criteria for later interpretation of field work.*

An example of a mafic-ultramafic layered intrusion is the Bushveld Complex, shown by the map in Fig. 10.5a: the intrusion is *ca.* 450 km in diameter and *ca.* 9 km in thickness. At the margins of the Bushveld, where it is emplaced into clastic sedimentary and volcanic lithologies, a 'border group' has been mapped, comprising thin sills of fine- to medium-grained gabbro, norite and pyroxenite. The vast bulk of the Complex is divided into four zones on the basis of rock-type and internal variations (Fig. 10.5). The *lower zone*, comprising thick, uniform orthopyroxenite layers, with interlayered olivine pyroxenites and harzburgites, is overlain by a layer termed the '*critical zone*' (so-called because of its content of chromium and hence its critical economic importance), containing interlayered norites and anorthosites near the top with anorthosite-norite-pyroxenite layers further down. Within the critical zone there are several prominent chromitite layers, each up to 1.5 m thick (Fig. 10.2), and also sulphide segregations, including a layer called the 'Merensky reef', which contains iron-nickel-copper sulphides and important platinum-group minerals. Above the critical zone is the *main zone* which comprises thick, rather uniform, layers of gabbro and norite, whereas the overlying *upper zone* contains olivine melagabbros, with magnetite-rich gabbros and anorthosite-magnetite layers near the top. A younger granite suite — thought to be unrelated, magmatically, to the Bushveld Complex — overlies and conceals the layered rocks in the centre of the outcrop (Fig. 10.5).

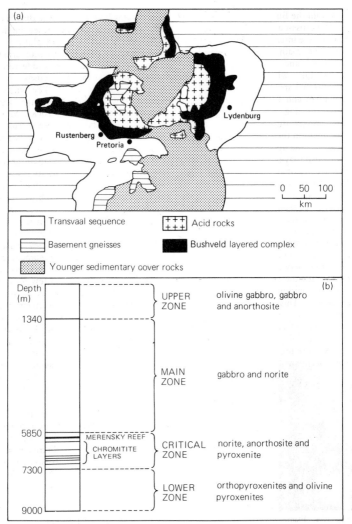

Fig. 10.5 (a) Geological map of the Bushveld region, South Africa, showing an arcuate outcrop pattern reflecting the saucer-shape of the mafic-ultramafic layered complex which is assumed to be present beneath the younger sedimentary outcrops. (b) Simplified vertical section through the Bushveld Complex showing the main features discussed in the text.

Within the Bushveld Complex there are few cross-cutting relationships and most of the interlayer contacts are planar. The main structure is one of laminated and rhythmically banded layers but with occasional post-depositional structures, such as flame structures and load casts of the kind shown in Fig. 10.6 (cf. Fig. 3.15). Interestingly, these structures are most prominent in the critical zone and near the top of the upper zone where major anorthosite layers enter the sequence. The density of anorthosite is much less than those of gabbro, norite and pyroxenite, even when these rocks are partially consolidated crystal mushes and are susceptible to deformation; hence, the prominence of penetrative structures, where anorthosite occurs, is almost certainly related to density inversion between the layers, compared with the more uniform density gradient within the gabbro–norite sequences. This correlation of post-depositional structures with the occurrence of anorthosite layers provides a useful guide when examining and interpreting other intrusions.

Fig. 10.6 Flame structures in a layered igneous sequence, in this case indicating that the more felsic material was of lower density than the more mafic material (compare with Fig. 3.15). Length of hammer is 25 cm.

On a smaller scale than the Bushveld and similar complexes, other mafic-ultramafic dykes, sills and plutons of all sizes occur in the continental crust. They may have been emplaced tectonically during metamorphism and thrusting, in which case they will be associated with sheared, deformed and high-grade metamorphic rocks. More simply, they may have been intruded as part of a larger igneous province (e.g. the Cuillin gabbros of Skye in Fig. 9.1b). Various combinations of the mafic-ultramafic rock-types described earlier are known: some intrusions are homogeneous gabbros, peridotites or even dunites, others show igneous lamination, and many are complex, multiple intrusions with mappable internal boundaries. Small intrusions of dunite are sometimes found in high grade gneiss zones where they may be only a few tens of metres to a few kilometres in size, elongated parallel to the surrounding foliation and may, themselves, show syn-tectonic folding and foliation. They may be extensively serpentinized (Fig. 10.1), particularly around their margins and, as well as serpentine, may also contain a range of secondary minerals such as chlorite, talc and magnesite.

Finally, it should be noted that the coarse-grained micaceous peridotites, known as kimberlites, that occur in deeply-eroded volcanic diatremes are classified as ultramafic intrusive rocks. These are considered in detail in Section 7.2.

10.3 Ophiolite complexes

Many tectonically-emplaced mafic-ultramafic bodies have an origin as slices of oceanic lithosphere. Such bodies are termed *ophiolites* and are preserved as a result of thrusting and/or uplift associated with plate tectonic processes. Since these ophiolite complexes have assumed great importance in interpreting the tectonic history of a region, it is useful to explain the field and petrological characteristics that enable them to be recognized. According to the Geological Society of America's Penrose Conference on ophiolites in 1972, an ophiolite is defined merely as an assemblage of mafic and ultramafic rocks, ideally with a stratigraphy as in Fig. 10.7. Thus a *complete* ophiolite sequence would provide representative samples of all the layers inferred to comprise oceanic lithosphere, but most mafic and ultramafic bodies with characteristics of ophiolites do not form complete sequences, possibly as a result of disruption during tectonic emplacement. However, once several of the following characteristics have been found in the field, some confidence may be placed in the identification of an ophiolite.

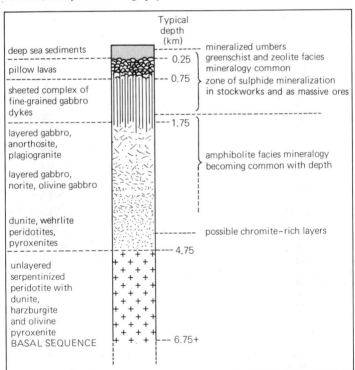

Fig. 10.7 Petrological section through a typical ophiolite sequence (further details in text).

First, ophiolite bodies are rarely found alone but tend to occur as intermittent elongate bodies aligned along a zone of major thrusting, such as the Alps and Himalayas. Within each ophiolite body, the rock-types often become progressively more ultramafic towards the base (remember, though, that the sequence may be dipping or even inverted due to tectonic emplacement). Ideally, the base may be seen to rest on a thrust zone which may be identified in the field by the presence of deformed and fragmented ultramafic rocks within the ophiolite, overlying, usually with sharp petrological and structural contrast, the local crustal rocks.

The *basal sequence*, itself the most variable in thickness, ranging up to 10 km normal to the dip of the base, may comprise generally unlayered mixtures of serpentinized peridotites, usually with strongly-deformed, 'tectonized' fabrics, but sometimes including zones of recognizable dunite and harzburgite. These rocks are interpreted as a '*mantle sequence*' from which the partial melts that formed the overlying (ocean crustal) layers have been extracted.

Overlying the mantle sequence, there are usually several kilometres' thickness of *layered mafic-ultramafic rocks*, resembling those of layered lopoliths, (e.g. Fig. 10.5). Near the base, cyclic successions of dunites, wehrlites and other peridotites or pyroxenites may occur and there is often evidence that repeated influxes of magma at different levels within the layered sequence each generated a cyclic unit. Thus, whereas the majority of the layered sequence comprises interlayered gabbros, norites and olivine gabbros, with increasing amounts of leucogabbro and anorthositic material towards the top of the layered sequence, parts of the sequence may be repeated due to multiple intrusion. The white 'anorthositic' rocks of ophiolites are usually rich in sodium-rich plagioclase feldspar and some of them contain substantial amounts of quartz (up to 30%), so that the term *plagiogranite* is more appropriate. The layered sequence is usually interpreted as representing an ocean crustal magma chamber which, in the same way as it received successive magma batches, gave rise to the overlying dykes and lavas by venting and emplacement of magma from its upper surface.

Fig. 10.8 Alternating bands of anorthosite (white), leucogabbro (pale grey) and mela-gabbro (dark grey) from the Oman ophiolite. Length of hammer is 1m.

The top of the layered sequence is marked by a decreasing proportion of coarse-grained rocks and the appearance of medium- to fine-grained dyke rocks trending approximately at right angles to the layering below. Higher in the ophiolite sequence, these dykes rapidly become a *sheeted complex*, in which almost all the outcrop may be occupied by interpenetrating *ca.* 1 m thick sheets of fine-grained gabbro (dolerite) or basalt (Fig. 10.9). The process may continue for some time at one focus of injection as shown schematically in Fig. 10.10. The sheeted complex of dykes may be *ca.* 1 km thick before it gives way upwards to a series of *pillow lavas*, typically *ca.* 0.5 km thick, though feeder dykes may continue through much of the lava sequence. Basalts predominate in the pillow lava sequence (e.g. Fig. 6.5) though more mafic picrites or more acid andesites and dacites may occur. Thick non-pillowed flows and occasional sills may also be found in the sequence. The uppermost lavas may be intercalated with various types of deep-sea sediments, ranging through cherts, shales and mudstones to occasional red-brown ferromanganoan hydroxide sediments, known as *umbers*. A thin sequence comprising purely sedimentary rocks is sometimes preserved at the top of an ophiolite body.

Fig. 10.9 Sheeted dyke complex from the Cyprus ophiolite (right) cutting upwards through pillow lavas (left).

Finally, the top 3–4 km of an ophiolite sequence is likely to show the effects of hydrothermal metamorphism due to the penetration of magmatically-heated sea water through cracks and fissures when the ocean crust was newly formed (see Fig. 10.7). The pillow lavas and dykes of ophiolites frequently show minerals of the zeolite and greenschist facies of metamorphism (see Fig. 12.1 for definition of these facies), whereas other dykes, lower in the sequence, and the layered gabbroic rocks, may

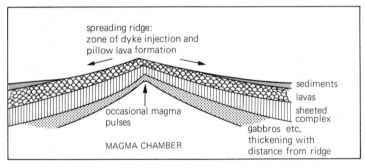

Fig. 10.10 Cross-section through an ocean ridge, showing (schematically) the processes that are envisaged to occur as pulses of magma rise to be emplaced as dykes or extruded as pillow lavas. Sea-floor spreading carries the newly formed dyke–lava sequence progressively further away from the ridge with time.

show amphibolite facies minerals. This means that, on a mineralogical scale, dark-green amphibole, originally at deeper levels, and pale-green chlorite, originally at shallower levels, should be recognized — both of which grew at the expense of pyroxene. White zeolite minerals may also be found in cavities or, more rarely, replacing feldspars in the dyke–pillow lava sequence (Fig. 10.7). During hydrothermal metamorphism, in general, it is common for base metals to be remobilized and concentrated into ore deposits (see Sections 3.2 and 8.3), and the upper regions of ophiolite sequences may contain *stockworks* along which mineralizing fluids were able to penetrate and which are often lined with chalcopyrite and other sulphide minrals. Where a major fracture focused the flow of mineralizing fluids into a sea-bed discharge zone, reaction between the hot, metal-enriched fluids and sea-water may have caused massive 'exhalative' sulphide ore bodies to be precipitated at the upper surface of the ocean crust; such sulphide ore bodies are also known from ophiolites. More commonly, however, the metals were precipitated as Fe-Mn-rich hydroxide umbers, carrying lesser amounts of copper, lead, zinc and other transition metals. So ophiolites can be economically important, not only for their deep chromite layers, but also for their high-level stockworks, massive sulphide mineralization and the related sedimentary umbers.

10.4 Summary of the field characteristics of mafic-ultramafic intrusions

1 Medium- to coarse-grained gabbro (cpx > opx), norite (with opx < cpx) and ultramafic rocks occur as isolated within-plate intrusions, and also as part of ophiolite sequences that are tectonically emplaced into continental suture zones.

2 Large mafic-ultramafic intrusions often display excellent igneous lamination. Individual layers may range petrologically from the extremes of anorthosite and economically-important chromitite, through a wide variety of peridotites, pyroxenites and gabbros; *the layers should be described in terms of grain size, mineral modality and texture* (cf. Figs. 10.3, 10.5). Penetrative, disruptive and 'sedimentary' slump structures are frequently displayed by the layering, especially where strong density contrasts (e.g. between melagabbro and anorthosite) exist between the alternating layers. In general, large mafic-ultramafic intrusions become more felsic upwards, and residual liquids from their fractional crystallization may evolve towards granitic compositions.

3 Most ophiolite complexes are igneous masses of elongate outcrop caught within known continental sutures. They are generally most ultramafic towards the base (cf. Fig. 10.7), and from the bottom upwards should comprise combinations of:

(i) a 'mantle' sequence of deformed dunite, harzburgite and pyroxenite,

(ii) a layered mafic-ultramafic sequence that becomes more leucocratic upwards,

(iii) a sheeted complex of dolerite dykes, and

(iv) basaltic pillow lavas with Fe-Mn umbers.

The effects of hydrothermal metamorphism should be recognized in the uppermost 3–4 km; high-level mineralized stockworks and, less frequently, massive exhalative sulphide ore bodies may also be found.

4   In addition to consideration of the points made above, field studies of mafic-ultramafic bodies will require reference to earlier chapters in this guide, in particular, the list in Section 8.6 provides pointers for use in the field.

Plutonic rocks IV: anorthositic and charnockitic associations

11.1 Anorthositic associations

Anorthosites are leucocratic crystalline rocks which are often coarse-grained (1–10 cm, exceptionally up to 1 m crystal size) and may occur as part of mafic-ultramafic sequences or as homogeneous intrusions (see below).

They are the most widespread igneous rocks to be dominated by a single mineral — plagioclase feldspar, which may be white (e.g. Fig. 11.1), light-grey or pale-brown in colour. Less commonly, anorthosites are dark-blue grey or even purplish brown and these types often show an iridescent sheen

Fig. 11.1 Many examples of anorthositic rocks are very striking in their appearance. This example of anorthosite (white) above amphibolite (dark grey) is from Nordfjord, central Norway. The contact is displaced *ca.* 5 m by a fault.

under bright light due to reflections from fine, multiple twin planes and Fe oxide inclusions within the feldspar (see Table 4.6) — this is the *schiller effect*. At first sight, anorthosites might easily be confused with carbonate rocks (e.g. marble and carbonatite, Chapter 9) or quartzite but, in the first case, hardness is distinctive, because that of feldspar exceeds that of calcite or dolomite, and in the second, cleavage and twinning may usually be identified in the plagioclase crystals forming anorthosite (whereas quartz has neither cleavage or twinning). Anorthosites may contain up to about 10% of other minerals, typically pyroxene and occasionally olivine, and this means that commonly related rocks with greater amounts of mafic minerals are *leucogabbros* (synonymous with anorthositic gabbros, Fig. 4.10), *gabbros* and *norites* (cf. Chapter 10). Anorthosites almost always contain a few per cent of Fe-Ti oxide minerals (magnetite and ilmenite) and occasional larger, commercially-exploitable concentrations of these minerals occur as layered segregations in some anorthosites.

Anorthosites occur as bands or lenses within larger mafic-ultramafic bodies (Chapter 10) and also as fairly uniform anorthosite intrusions having horizontal dimensions comparable with batholiths but believed (mainly

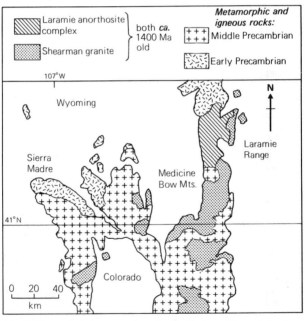

Fig. 11.2 Outline geological map to show the distribution of Precambrian igneous and metamorphic rocks across the Colorado–Wyoming border area.

from geophysical data) to have a sheet-like form, no more than a few thousand metres in the vertical dimension. Many of these large anorthosite massifs are associated with the mid-to-late Precambrian (Proterozoic) alkaline complexes discussed in Chapter 9, which contain alkali feldspar granites and syenites. For example, the two suites of *ca.* 1400 Ma old rocks represented in Fig. 11.2 are the Shearman alkali granite-granite complex and the Laramie anorthosite-syenite complex. Detailed field work on anorthosite complexes may therefore involve examination of a range of granitoid rock-types.

A closer inspection of anorthosite massifs usually shows them to be multiple intrusions. A particularly well-studied example is the Rogaland complex of Southwest Norway (Fig. 11.3) which contains both syn- and post-tectonic anorthosite intrusive components. More significantly, although anorthosites predominate, several other related coarse-grained rocks may occur in these complexes. Perhaps

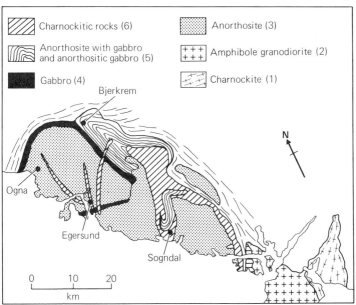

Fig. 11.3 Simplified geological map of the *ca.* 1000 Ma old Rogaland igneous complex of southwest Norway. An ancient pre-tectonic anorthosite-gabbro body (rock-types 3 and 4: the Ogna-Egersund massif) is cut by a complex syn-tectonic anorthosite-gabbro-charnockite layered lopolith (rock-types 5 and 6: the Bjerkrem-Sogndal body) and other anorthosites (again rock-type 5) to the south. Related charnockite and amphibole-bearing granodiorite intrusions (rock-types 1 and 2) also occur to the south. Dashed lines indicate structural trends in basement gneisses.

most commonly, parts of an anorthosite body may grade through anorthositic gabbro (65–85% plagioclase,
15–35% mafic minerals) to norite (20–
65% plagioclase, 35–80% mafic minerals, mainly orthopyroxene); note that
norites rather than gabbros are common in these complexes.

11.2 Charnockitic associations

Besides biotite (or amphibole) granites
and alkali feldspar granites, another
minor group of silica-rich rocks,
known as *charnockites*, form plutonic
complexes, sometimes in association
with anorthosites (cf. Fig. 11.3). In the
literature, these are loosely defined as
orthopyroxene-bearing rocks of
diorite to granite composition, i.e.
they are quartz-feldspar orthopyroxene-bearing rocks (which
sometimes contain brown iron-rich
olivine, fayalite). One of the problems
in recognizing charnockites is that frequently they are deceptively dark-
coloured, containing bluish-grey
quartz and dark-grey feldspars. Careful study of the rock texture under a
handlens should enable dark-coloured
quartz to be identified. The presence
of bronze orthopyroxene crystals,
showing right-angled cleavages, also
often helps with charnockite identification. The mineral fabric of charnockites often shows preferred mineral
alignment due to the probable origin
of these rocks by high-pressure metamorphism of normal granodiorites and
granites in the high pressure–
temperature *granulite* facies (see Fig.
12.1).

11.3 Summary of the field charac-
teristics of anorthositic and charnockitic intrusions

1 Anorthosites contain 90% or more
   plagioclase feldspar, which is
   commonly white, pale grey or
   brown, together with small
   amounts of pyroxene, olivine or
   Fe-Ti oxide. They occur as layers
   within larger mafic bodies, or as
   sheet-like intrusions in association
   with alkaline complexes.

2 Charnockites are quartz–feldspar-
   orthopyroxene rocks usually produced by high grade metamorphism of normal granites and
   granodiorites; in the field, the
   original intrusions therefore appear
   strongly deformed.

Reference has been made at various points in this field guide to the relationship between igneous rocks and regional metamorphic fabrics (e.g. the terminology introduced in Fig. 8.8), to the kinds of complex that result from repeated phases of metamorphism and intrusion (e.g. Fig. 8.9) and to some of the rock-types produced by regional metamorphism of igneous rocks (e.g. charnockitic granites resulting from granulite facies metamorphism). This chapter provides a brief summary of the main rock-types that are encountered (a) when igneous rocks are metamorphosed and (b) in zones of 'ultra-metamorphism' where igneous and metamorphic rocks are literally 'mixed' and interlaminated to form *migmatite* complexes. Further details of the terminology and rock-types introduced here can be found by consulting the further reading list, and the complementary field guide to metamorphic rocks (Fry, 1984).

Metamorphosed igneous rocks may be recognized and distinguished from metamorphosed sedimentary rocks by (a) having an overall mineral composition corresponding to igneous rocks as described earlier, (b) preservation of igneous textures (Chapter 4) and (c) recognition of igneous field relationships such as cross-cutting veins and contacts. Metamorphism of igneous rocks takes place under a range of conditions from low pressure and temperature (*low-grade*) to high pressure and temperature (*high-grade*) (Fry, 1984, Section 5.5). Under conditions of low-grade metamorphism, igneous rocks may retain the chemical composition of the original rock (except for the addition of H_2O and CO_2), the textures may be slightly obscured by alteration and shearing (cf. Fig. 8.4) of the igneous minerals (e.g. conversion of mafic minerals to chlorite, and of feldspars to epidote and white mica), and igneous field relationships may be clearly preserved. However, with increasing metamorphic grade, textures and field relationships become progressively obscured, so that high-grade metamorphism may lead to complete recrystallization of the original igneous texture, loss of original field relationships and possible changes in chemical composition beyond the loss of H_2O (termed *metasomatism*). Further information on the metamorphism of igneous rocks is given in Fry (1984).

12.1 Metamorphism of igneous rocks

First, some definitions of textures that may apply to igneous rocks which have undergone metamorphism:

Foliation describes a planar structure recognizable in hand specimens where metamorphic minerals, usually micas, have grown at right angles to the dominant stress direction

Rocks in which micaceous foliation is pronounced are termed *schists,* and the well-developed panar foliated texture is known as *schistosity.* A coarse-grained banded rock in which there is irregular foliation with alternating bands of granular felsic and mafic minerals is known as a *gneiss.* A particular distinctive texture in metamorphosed granitoid rocks occurs where micaceous foliation is wrapped around centimetre sized pools of *augen* (German 'eyes'; crystals of feldspar with 'tails' parallel to the foliation), and is known as *augen gneiss* (see 12.5f.).

When describing metamorphosed igneous rocks in the field it is important to:

(i) *identify the igneous and metamorphic minerals present,* the former to aid recognition of the original igneous rock-type and the latter to evaluate the grade of metamorphism. Bear in mind that, although metamorphic minerals may grow over original igneous textures, they may leave *pseudomorphs* of the original minerals (see detailed notes in Fry, 1984, Chapter 7). Note also that retrograde metamorphic reactions (e.g. amphibolitized basalt retrograded to talc-chlorite schist) may yield an artifically low estimate of metamorphic grade. Fortunately, retrograded metamorphic mineral assemblages are rarely uniformly dispersed throughout an igneous body but are concentrated in zones, such as *shear zones,* where hydrothermal fluids were most able to affect the rock. Careful studies may

allow non-retrogressed assemblages to be recognized away from intensely-altered zones.

(ii) *describe accurately the rock texture at both hand specimen and outcrop scale* noting, in particular, the composition, thickness, orientation and state of deformation of gneissose banding for correlation with structural studies.

(iii) *make sketches and/or take oriented photographs for future reference,* especially where microfabrics and mineralogy are to be studied subsequently in thin section.

A combination of petrological and structural studies from sampled outcrops over a complex metamorphosed igneous body may allow a full picture of the thermal and burial history of the body to be defined in terms of several phases of deformation, each of which may be related to new mineral growth. A full description of such studies lies outside the scope of this handbook and, for more information, reference should be made to the further reading list.

Figure 12.1 is a metamorphic facies diagram which is used to describe the suites of mineral assemblages that result from changes induced by burial and heating. The facies given in this diagram were originally defined using the assemblages found in metamorphosed mafic and ultramafic igneous rocks — basalts and gabbros — which tend to preserve their original grain size as new minerals grow during metamorphism. Olivines, pyroxenes and calcium-rich plagioclases are repaced by serpentine, chlorite, epidote and mica (in the *greenschist* facies, Fig. 12.2) and by hornblende (in the *amphibolite* facies) with increasing pressure and temperature. High

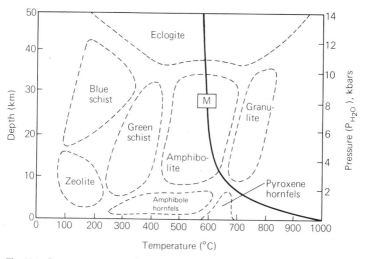

Fig. 12.1 Pressure–temperature diagram showing the fields of stability for different metamorphic mineral assemblages (facies), described in the text and Table 12.1, in relation to the beginning of melting curve (*M*) for water-saturated crustal rocks. (cf. Fry, 1984, Fig. 5.5)

Fig. 12.2 Serpentinized layered mafic-ultramafic sequence showing shear deformation and faulting (top right to bottom left, roughly parallel to the pen) from the Lizard Complex, Cornwall. The sequence underwent regional metamorphism in the greenschist to lower amphibolite facies and was then retrograded to a serpentine-talc-chlorite assemblage.

pressure–low temperature metamorphism gives rise to *blueschists*, which are characteristically dark blue amphibole and pyroxene-rich rocks, and to *eclogites*, which contain blue-green pyroxene and spectacular deep-red garnet. Low pressure–high temperature metamorphism produces finely banded *hornfels* (Fig. 8.14b). Table 12.1 summarizes the main features of these rock-types, as viewed in hand specimen, and should be used for field identification purposes.

In contrast to the minerals of metamorphosed basic rocks, the quartz and alkali feldspars of relatively acid rocks are less susceptible to metamorphic reaction and, apart from the hydrothermal production of white micas and clays, little effect may be noticed until high-grade conditions are encountered. High pressure–temperature recrystallization may then give rise to gneissose fabrics (e.g. the diorite and granodiorite gneisses in Fig. 12.3 a,b). Usually these are medium- to coarse-grained banded metamorphic rocks, in many of which alternating light 'granitic' and darker biotite-amphibole-rich bands vary in width from centimetres to tens of metres (e.g. Fig. 12.4). Although many

Table 12.1 Characteristic features of metabasic rocks in different metamorphic facies

Rock-type	Colour, grain size and texture	Mineralogy
Greenschist	Pale- to medium-green, usually fine-grained; schistose texture (i.e. well foliated with micaceous layers, often giving a silky appearance)	Chlorite, epidote, white mica, sodium-rich plagioclase, occasional amphibole (actinolite needles or tabular hornblende)
Amphibolite	Dark-green, greyish green or black; medium- to coarse-grained; occasionally foliated and often showing preferred orientation of hornblende crystals	Hornblende dominant, minor plagioclase feldspar, occasional pale-green epidote and red garnet
Blueschist	Medium- to dark-blue to bluish grey; medium- to coarse-grained; granular and sometimes foliated	Blue soda amphibole, chlorite, green jadeitic pyroxenes occasionally replacing sodium-rich feldspars
Eclogite	Spectacular mid- to dark-green rock, often with spots of red mineral grains; medium- to coarse-grained; crudely foliated or granular texture	Green jadeitic pyroxene deep-red garnet and occasional blue kyanite
Hornfels	Mid- to dark-grey finely banded rock; fine grain size usual; granular appearance, non-foliated	Segregated bands of pyroxene and feldspar; occasional mica and needles of andalusite

gneisses (so-called *paragneisses*) may be shown from field association with

(a)

(b)

Fig. 12.3 (a) Diorite gneiss from the French Alps, showing prominent micaceous foliation parallel to the knife and the development of feldspar augen (most prominent in top centre). (b) Strongly folded granodiorite gneiss from the Rocky Mountains, Colorado, showing fine mafic-felsic segregations. Although the bands may be roughly parallel, the felsic segregations are discontinuous and, in three dimensions, are deformed (squashed) tabular pods. Pen is 15 cm long.

other metasedimentary strata, and preservation of sedimentary structures, to have a sedimentary origin, most gneisses (*orthogneisses*) are believed to be derived from igneous rocks. These take their names from the mineral composition of the parental rock which appears in the gneiss itself (e.g. diorite gneiss, granodiorite gneiss, etc.).

Fig. 12.4 Garnetiferous granite gneiss from Rio de Janeiro, Brazil. The leucocratic bands comprise quartz, feldspar and red euhedral garnets (dark equidimensional grains); the melanocratic bands contain biotite, sillimanite and feldspar. The metamorphic minerals, sillimanite and garnet, indicate gneiss formation at high pressure–temperature conditions, at least in the high pressure part of the granulite facies field of Fig. 12.1.

12.2 Migmatite complexes

There is a large area of overlap between igneous and metamorphic rocks, and this is particularly evident among coarse-grained leucocratic rocks. Migmatites fall across this boundary and literally contain a mixture of the features characteristic of

145

both major rock groups. *By definition, migmatites are composite rocks containing deformed parent igneous or country rock* (the palaeosome) *intimately interbanded with newer igneous material (*the neosome*); this is usually dominated by quartz and/or feldspars* (the leucosome), *but occasionally is mafic* (the melanosome). They often have gneissose fabrics, but unlike simple gneisses, migmatites usually show clear evidence of igneous material having been introduced or, at least, of having been produced by partial melting *in situ* whilst earlier fabrics were preserved. The leucosome may take the form of veins or layers, or simply appear to provide a host for xenolith-like inclusions of palaeosome (see Fig. 12.5, which is discussed below). Further evidence of the introduction of melt, produced under ultrametamorphic conditions (syntectonic melting) lies in the fact that some migmatites can be shown to grade laterally into rocks of true igneous, often granitoid aspect, through the progressive loss of the palaeosome. For this reason, and the fact that migmatite injection gneisses are common in the marginal zones of deeply-eroded granitoid batholiths, it is common to think of them as representing sub-batholith melt zones from which magmas were incompletely extracted to form major intrusions. Migmatites are also often associated with granulite facies gneisses, comprising high-temperature pyroxenes and calcium-rich plagioclases, which may represent refractory residues remaining after partial melting and liquid extraction.

Figure 12.5 illustrates schematically the characteristic appearance of some common types of migmatite and should be used for comparisons in the

field when describing examples at outcrop. Migmatites displaying *raft structure* (Fig. 12.5a) are a natural extension of net-vein complexes (Fig. 3.9b) in which the xenoliths of the palaeosome have become detached and have floated like rafts in the invading magma. A distinctive feature is that the inclusion fabrics are chaotically oriented, indicating rotation with respect to one another; they may be sheared and may show diffuse boundaries indicating dissolution into the magma. *Vein-structured* migmatites (Fig. 12.5b) comprise irregular elongate pods and vein-like leucosomes that have been injected as tabular forms into a system of shear fractures in the palaeosome. Note that the orientation and form of the veins bears no relationship to fabrics, such as foliation, in the palaeosome, and is usually discordant to such fabrics. This distinguishes vein-structured migmatites from highly deformed granodiorite gneisses (e.g. Fig. 12.3b), in which the leucocratic bands follow the gneissose structure — but see also comments below on fold structured migmatites. Occasionally, the leucosomes of vein-structured migmatites occupy a more regular system of layers that apparently were injected along planes in the palaeosome — these are termed *layered-structured* migmatites or, occasionally, *lit-par-lit injection gneisses.*

Figure 12.5c illustrates a dilation-structured migmatite in which the leucosome magma invaded the fracture spaces between *boudinaged* (stretched and broken) blocks of a competent layer in the dilated palaeosome. Pinch and swell structures in less-competent layers that flow rather than fracture are also common in this type of migmatite. Similar invasion into fracture spaces in folded, competent palaeosome gives rise to *fold-structured* migmatites (Fig. 12.5d) and various combinations of complex multiple folding and of syn- or post-tectonic injection may be recognized. Flow structures in the leucosome may be broadly parallel to the palaeosome foliation (as shown in Fig. 12.5d) but careful examination will always reveal discordant zones, indicating magma injection into a pre-

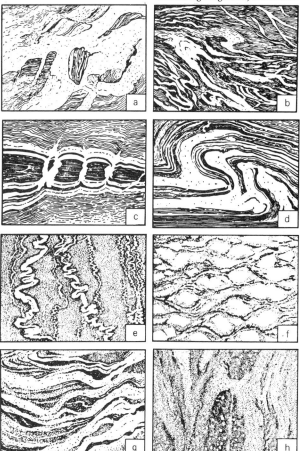

Fig. 12.5 Schematic illustration of some important migmatite structures described in the text: (a) raft structure; (b) vein structure; (c) dilation structure; (d) folded structure; (e) ptygmatic structure; (f) augen structure; (g) schlieren structure; (h) nebulitic structure.

existing host as distinct from the development of leucocratic bands during syn-tectonic gneiss formation.

In *ptygmatic-structured* migmatites (Fig. 12.5e), the leucosome forms tightly folded vein-like segregations, or 'ptygmae' (Greek, 'folded matter'), comprising equigranular, coarse-grained aggregates of quartz and feldspars. The veins are discordant to structures in the palaeosome but the fold limbs are often parallel to schistosity in the host, implying that shear movements along schistosity planes may play some part in their development. However, the occurrence of ptygmae in palaeosomes without schistosity indicates that their origin may be more complex (a summary of hypotheses appears in Mehnert, 1968). *Augen structure* in migmatites (Fig. 12.5f) usually comprises leucosome pods of feldspar developed within an earlier palaeosome fabric which has been displaced by crystal growth. A distinction is made between true augen-structured migmatites, in which the palaeosome contains relict metamorphic structures, and augen gneisses (Section 12.1), in which the feldspars segregated during a metamorphism that produced a simple continuous gneissose structure. In practice, this distinction can be made in the field only with careful attention to the details of fabric and texture.

Migmatites in which the mechanical mobility of both constituents was greater than that of the examples previously described may form *schlieren-like structures* (Fig. 12.5g), in which the palaeosome inclusions are tapered and twisted, and show new metamorphic textures developed during migmatization. The pressure-temperature regime in which such migmatites developed was, clearly, much higher than for types in which earlier palaeosome structures are easily recognized (Fig. 12.5a–f). Taken to its logical conclusion, schlieren give way to *nebulitic-structured migmatites* (Fig. 12.5h), in which the palaeosome and leucosome can no longer be separately identified, although the 'ghost' structures are still preserved. Further detailed explanation of all the migmatite structures described here can be found in Mehnert's (1968) book on this subject.

Migmatites are widespread in the Middle to Late Precambrian terrain of many continental interiors; particularly well-known examples occur in Finland where migmatites, granites and granite gneisses comprise some 70% of the land area. Finally, a note of caution: it is perhaps wise to visit a large number of outcrops before describing a rock-type as a definite migmatite. There are many examples of deformed granitoid rocks that are not truly migmatitic in the 'mixed' sense of the word. This is why the term is best used in a regional context, when firm proof of melt mobilization accompanying or post-dating metamorphic deformation has been recorded. Hence, the term *migmatite complex* is preferred for field use and the term *migmatite* for an individual rock specimen.

References and further reading

Books containing sections on relevant practical techniques for field studies

BARNES, J.W. (1981) *Basic Geological Mapping*, Geological Society Handbook, The Open University Press, Milton Keynes/Halstead Press, New York/Toronto, 112 pp.

COX, K.G., PRICE, N.B. & HARTE, B (1974) *The Practical Study of Crystals, Minerals and Rocks*, London, McGraw-Hill, 245 pp.

DIETRICH, R.V. & SKINNER, B.J. (1979) *Rocks and Rock Minerals*, Chichester/New York/Brisbane/Toronto, Wiley, 319 pp.

FRY, N. (1984) *The Field Description of Metamorphic Rocks*, Geological Society Handbook, The Open University Press, Milton Keynes/Halstead Press, New York/Toronto, 110 pp.

LAHEE, F.H. (1918) *Field Geology* New York/London/Toronto, McGraw-Hill, 926 pp.

MOSELEY, F. (1981) *Methods in Field Geology*, Oxford/San Francisco, W.H. Freeman, 211 pp.

TUCKER, M. (1982) *The Field Description of Sedimentary Rocks*, Geological Society Handbook, The Open University Press, Milton Keynes/Halstead Press, New York/Toronto, 112 pp.

General texts concerned with igneous rocks

COLEMAN, R.G. (1977) *Ophiolites: Ancient Oceanic Lithosphere?* Heidelberg, Springer-Verlag, 229 pp.

COX, K.G., BELL, J.D. & PANKHURST, R.J. (1979) *The Interpretation of Igneous Rocks*, London, George, Allen and Unwin, 450 pp.

DIDIER, J. (1973) *Granites and Their Enclaves: the Bearing of Enclaves on the Origin of Granites*, Amsterdam/London/New York, Elsevier Scientific, 393 pp.

EVANS, A.M. (ed.) (1981) *Metallization Associated with Acid Magmatism*, Chichester/New York/Brisbane/Toronto, Wiley, 385 pp.

IRVINE, T.N. (1982) 'Terminology for Layered Intrusions', *J. Petrol.* 23, 127–162.

MACDONALD, G.A. (1972) *Volcanoes*, New Jersey, Prentice-Hall, 510 pp.

MEHNERT, K.R. (1968) *Migmatites and the origin of granitic rocks*, Amsterdam, Elsevier, 393 pp.

WILLIAMS, H. & McBIRNEY, A.R. (1979) *Volcanology*, San Francisco, Freeman, Cooper & Co., 397 pp.

Index

153